The Complete Price Guide To Antique Radios:

# Tabletop Radios

## Volume 4

The Complete Price Guide To Antique Radios:

# Tabletop Radios

## Volume 4

By Mark V. Stein

Radiomania® Books 2109 Carterdale Road Baltimore, Maryland 21209 U.S.A

Website: www.radiomania.com E-Mail: rpbook@bellatlantic.net Fax: 800-891-4484

Published by
Radiomania® Books


ISBN: 0-9647953-6-1
Printed in The United States of America

BOOKS BY MARK V. STEIN

***The Complete Price Guide to Antique Radios Series:***

*Tabletop Radios*, Volume 1*
*Tabletop Radios*, Volume 2*
*Tabletop Radios*, Volume 3*
*Tabletop Radios*, Volume 4*
*Pre-War Consoles*
*The Sears Silvertone Catalogs: 1930 - 1942*
*Tabletop Radios series originally titled 'Machine Age to Jet Age'

***The Collector's Guide to 20th Century Modern Clocks:***

*Volume 1 - Desk, Shelf and Decorative*
*Volume 2 - Wall and Mantle (2004)*

Further information regarding purchase of this and other titles published by Radiomania Books, refer to the final pages of this book or visit the Radiomania website:

**www.radiomania.com**

Note that all books purchased directly from Radiomania are personally signed by the author and shipped free of charge within the United States.

**NOTICE:**

**The market values indicated in this reference are based on a number of different sources. Actual prices will vary dependent on many variables. Neither the author nor the publisher assumes responsibility for losses which might result from the use of this book.**

*Cover design by Jane Rubini*
*Book design by Mark V. Stein and Jane E. Rubini*

# TABLE OF CONTENTS

## TABLE OF CONTENTS

## TABLE OF CONTENTS

## TABLE OF CONTENTS

# INTRODUCTION

I am pleased to present to you the new fourth volume of *The Complete Price Guide to Antique Radios: Tabletop Radios*, previously titled *Machine Age to Jet Age*. This reference is comprised of all new and expanded listings not appearing in prior volumes. The total number of models now cataloged in this four volume series is well over 12,000.

Our listing details have continued to be refined and expanded. The format now includes manufacturer, model number and year, tube and band counts, color, materials and market value variations. Also described are options such as bandspread or motorized tuning. Additionally, this volume utilizes easily recognizable symbols to identify specific high profile characteristics. These include world's fair commemoratives, Ingraham cabinets and industrial designers. Canadian radios are also now noted with a specific icon.

Based upon collectors' enthusiastic response, our references have continued to be expanded and updated. Watch for new titles and price guides from Radiomania Books. Visit our website at:

**www.radiomania.com**

There you can find special internet book offers and register to be the first notified of new book releases and promotions.

Thank you for your continued support!

Mark V. Stein
Author and Publisher
Radiomania® Books

# PRICE GUIDE VALUES

In compiling this book the question as to whether to include prices at all was one of major concern to both the author and contributors. As most collectors are aware, even the best price guide, if not initially flawed, is soon obsolete. The range of prices paid for most items is wide. One can pay anywhere from a few dollars to a few hundred for a given item, dependent on where it was purchased and from whom. Thus the task of establishing a 'market value' is difficult if not impossible. This problem given, it was still generally felt that the value of including prices, at least as a general market gauge, would outweigh the disadvantages of omitting them completely.

In establishing values a number of sources were utilized including auction results, classified ads, meet pricing, collector valuations and the author's personal experience having been both a collector and dealer for over fifteen years. Prices in this book represent items in average condition. Average meaning that the radio is intact but 'as found'. No repairs would have been made, either electrically or otherwise. If wood, the finish would be original. It might show some wear, but would be presentable in a collection once cleaned up. If plastic, the cabinet would be free from cracks and chips. Plaskon radios in average condition might evidence some minor stress lines but nothing which would detract from their aesthetics. Values followed by an asterisk (*) symbol represent rarer items, only a few of which have traded hands. In such cases the value listed represents an estimate of what a collector might expect to pay for an average example. With regard to catalin radios, price will vary widely based on color, condition, marbling, and variations. This reference does not attempt to address the intricacies of valuing catalin radios, but merely to provide the reader with a baseline gauge.

A word about dealers and dealer prices: expect to pay a premium when purchasing from a dealer. The dealer offers one the luxury of eliminating the time consuming hunt through yard and estate sales, flea markets, antique shows and the like. It is he who goes through the trouble of rooting out those hard to find items. Ones which you might not happen upon except after years of hunting yourself. Dealers inventories represent long hours and related expenses and must reflect those additional costs.

# GENERAL CONSIDERATIONS

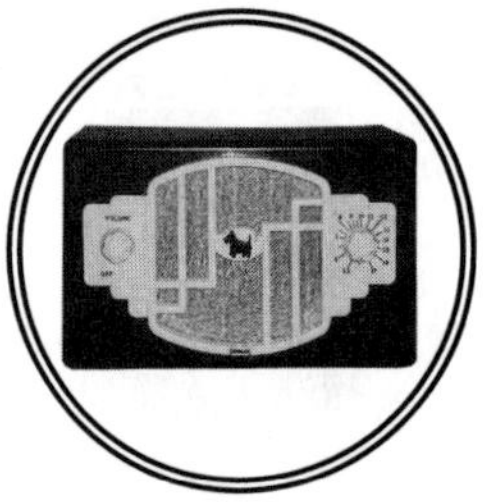

To assist the collector in determining the value of a given radio, here are some general rules of thumb, gained from the author's tenure as both a dealer and collector for the past two decades. Please remember, as with all rules, there are always exceptions. If you are uncertain about a specific item you will do best to ask another collector whom you trust or seek the expertise of a knowledgeable appraiser.

## DESIGN PEDIGREE

The term 'Design Pedigree' has appeared in recent years and is used when referring to an item (in this case a radio cabinet) whose design was drafted by a noteworthy industrial designer. Much attention has been paid to such mass produced items by collectors and curators. All such items command a significantly greater market and, therefore, higher price tag. Industrial designers including John Vassos, Gilbert Rohde, Henry Dreyfuss, Walter Dorwin Teague, Norman Bel Geddes and Russell Wright have all designed radio cabinets for major manufacturers. A collector's knowledge of Design Pedigree will give him a significant edge.

There are a host of radios which are attributed to industrial designers, but which are not documented with certainty. Many radio cabinet designs were not patented. Those design patent records which do exist are not well organized and research is difficult. Most radio 'pedigrees' are discovered through period advertising, industry periodicals and publications specific to industrial designers. Still, there are numerous assumed attributions which remain undocumented. Many are common knowledge. Some may be erroneous just the same. Most will never be proven either way.

This author has special interest in industrial design and his knowledge in this regard has been noted in this price guide with either the documented designer (Des.) or undocumented attribution to a designer (Attrib.) designation. If you have documented information relative to any radios in our price guides which has not been so noted, or otherwise, please contact the author via this publisher so that he may make appropriate notations in future books and updates.

## CABINET DESIGN

Short of a design pedigree, cabinet design remains of significant import. Striking art deco cabinetry, intricate repwood (also known as sirocco) panels, exotic veneers, marquetry and inlays are all of prime consideration in establishing value. These radios, although not attributed to a specific industrial designer, may be well known and highly valued among collectors.

## BRAND NAME

Among collectors, certain brands are desirable in and of themselves. Often times, the rationale might be the city in which the manufacturer was located. Examples of such brands include Detrola (Detroit), Simplex (Sandusky, OH), and several Los Angeles manufacturers including Troy, Herbert Horn/Tiffany Tone, Gilfillan, Packard-Bell and Jackson-Bell.

Other rationales for collecting a specific brand include quality (eg., Atwater Kent, Stromberg-Carlson, Zenith, EH Scott, McMurdo Silver) and design diversity (eg., General Television, Climax, Clinton). Many collectors focusing on a single brand compete furiously for new acquisitions to round out their collection.

## TUBE AND BAND COUNTS

Tube and band counts are both general indicators of a radio's position in the production line for a given year. Although single band radios were the rule until 1934, tube counts varied in these early years, generally with as few as four tubes and ranging to twelve or fifteen. The number of tubes determined the cost and complexity of the circuitry. More tubes indicated greater sensitivity, fidelity and power amplification. Beginning in 1934 two band radios (broadcast/shortwave) radios were introduced in mass production. In 1935 the shortwave bands were further divided into smaller band ranges. Tube counts were generally correlated with band counts from 1935 onward, with both increasing in number, the higher one went in a given manufacturer's product line. The higher the rank in the production line, the greater the cost to the consumer. This usually translated into lower production numbers. The number of high end radios surviving today is thus extremely limited.

High tube counts aren't always associated with high quality. Of historical note is the 'tube count scandal' of the 1930's. Until 1934/35 the public correctly associated higher tube counts with both greater value and quality. Soon, many less reputable and/or financially unstable manufacturers began adding more, unfunctional, tubes to their radios to increase sales. These additional tubes were often ballast (resistor) or other inexpensive tubes. Often times they were not connected to the main circuitry, but only provided with enough power to glow or heat up and give the consumer the impression they were performing some function. A six tube/three band radio was easily transformed into a seventeen tube/three band radio with little additional cost. The market became flooded with such radios, all selling for a fraction of those with legitimate tube counts and functionality. In 1937 this scandal was revealed to the public and the practice was brought to a halt. Particular pressure was applied from the radio trade industry itself. Several manufacturers were driven to close their doors, either through legal action or from diminished sales due to their loss of consumer trust, the most notable of these being Kadette. Beware of high tube count radios dating from 1935 through 1938. Again, most were off-brands and many may not be included in this reference.

## HIGH FIDELITY

High Fidelity (Hi-Fi) was introduced in the early 1930's and was provided by more intricate circuitry and increased amplification so as to significantly enhance the quality of sound reproduction. Only the ultra high-end radios produced prior to WWII offered this feature. Hi-Fi required a greater number of tubes and a higher level of sophistication in manufacture.

## BANDSPREAD TUNING

Bandspread tuning technology allowed the consumer to take a small portion of the wave band and fine tune by spreading this portion across the dial. A multi-band radio so equipped had significantly increased sensitivity. Few home radios were produced with this feature, which ultimately became most common in amateur (HAM) and shortwave/dx (long distance broadcast) receivers.

## MOTOR TUNING

Motor tuning was offered for a brief period, primarily 1937-1939, and utilized an electric motor attached to the tuning mechanism. With the push of a button or slight turn of a knob, the consumer could tune the radio to a specific preset station or scan the band. The motor tuning option was offered by several manufacturers, primarily in high-end radios .

## FREQUENCY MODULATION

Manufacturers began offering frequency modulation (FM) reception as an option beginning in 1940/41. The first FM receivers were at the very high end of their respective product lines. It is important to note that the pre-war FM band differed from that of today. A pre-war FM radio will not receive today's FM stations.

## POWER SOURCE

Most of rural America was not 'electrified' in the 1930's. Therefore, many radios were designed for operation with direct current power (DC). The power source might be one, two or three batteries (A, B & C Batteries). DC power was also provided motor generators and windmills (ie., Zenith's Wincharger). Input voltage also varied widely. When evaluating the worth of a vintage radio, the power source, AC vs. DC, is important. A DC powered radio will require modification or a converter if it is to be operational. Conversion cost ranges around $100 –150. Even with a converter, a DC powered radio is less desirable and, therefore, less valuable than an equivalent AC counterpart. This value differential is typically 25–50%.

# CHASSIS AND TRIM CONSIDERATIONS

In buying any radio it is important to visually inspect the chassis, particularly if you plan to restore the set to working order. Additionally, the absence of, or damage to, trim pieces is important. Here are some general guidelines to keep in mind:

## CHASSIS CORROSION

Minor chassis surface oxidation/corrosion is typical and acceptable to all but the most finicky of collectors and should not deter from the value. Extreme corrosion can be an indication that the radio was submerged at one time or at least has seen a lot of humidity. If so, the bulk of the internal components may need to be replaced. Even if you plan only to display the radio, such damage decreases its value by virtue of the fact that it is less desirable to most other collectors.

## RODENT DAMAGE

Most of us have, at some time, come across a radio the chassis of which has been stuffed with leaves, feathers, pieces of acorns and other unlikely items. Other radios may have exposed frayed exposed wiring or an unusual and unpleasant odor. Such phenomena are common, particularly when the radio has been stored in an outside barn or garage. Squirrels, raccoons, rats, mice, mudwasps and other assorted wildlife can be quite adaptive in making an abandoned radio their home. The damage caused by such occupancy can be severe. If you come across such a set and are tempted to buy it, inspect it carefully. The observed superficial problems are a likely indication of more extreme internal destruction. It is wise in such instances to take the trouble to remove the chassis for careful inspection.

## DAMAGED TRANSFORMER

A common problem found in AC powered radios, and one that may be costly to remedy, is that of the 'smoked' transformer. Caused by a short circuit or overload, usually from faulty aged internal electronic components, damage to the transformer necessitates its repair or replacement. Fortunately, such problems are often easy to spot. Look for smoke damage to the chassis and on the inside of the cabinet. Also look for black goo, which may have hardened over time, oozing from the transformer itself.

If the transformer is damaged, a replacement is the first option. New transformers are available from a number of sources, however their variety is limited. One might try to find a junker radio or salvage dealer with a transformer of matching specs. Otherwise, the last alternative is repair, a tedious process in which the transformer is dismantled and then rewound with copper wire. Transformer replacement can cost $20 – $100+ assuming you are lucky enough to find a replacement. Transformer repair is virtually unavailable today to the author's knowledge.

## MISSING TUBES

Missing tubes are generally easy to replace, particularly amongst radios from this era. The resources are numerous. The internet is an excellent resource for buying tested used tubes and there are many reputable dealers. For new and NOS (new old stock) tubes, check with local repair shops, mail order companies and specialty houses. It is a good idea to know your resources when considering the purchase of radio tubes. Some early tubes can cost as much as $50-75 each, used and testing good. Also be aware that there are components known as ballast tubes which resemble metal cased tubes superficially and plug into tube sockets, but which are actually resistors. From the author's experience, you can count on 50% of found ballast tubes needing replacement. Unlike filament tubes, few ballast tube NOS has survived and there are no known current resources for newly produced or good used components. Thus, replacing a failed ballast will require much perseverance or rewiring.

## KNOBS

If one or more knobs are missing from a radio its value will be diminished. The purchaser of such a radio must resolve himself to either completing the set or substituting another complete set that looks aesthetically correct. If you chose to complete the set with correct knobs, you can either search for a replacement at swap meets or through a salvage dealer at a cost of about $5 to $25 per knob, or purchase a reproduction, usually in the same price range. There is now a wide variety of reproduction vintage knob types available and stocked by many vendors, most are plastic castings, a few are turned wood. If you need a knob which is not available as reproduction, there are a few specialty vendors who will make a casting and reproduction from a sample knob provided. Cost for such knobs typically begins at $25. Regardless of how you decide to deal with the missing knob(s), the paying price for the radio should be reduced accordingly.

## DIAL LENSES, CRYSTALS AND SCALES

Covering the dial area on most radios made after 1933 is a glass or plastic dial lens. There area a few vendors who will reproduce a plastic lens from the damaged original or tracing of the opening, at a cost of around $20. Round glass dial lenses, typically made from convex round pieces of glass can usually be replaced with a clock crystal of the same diameter. Check with local clockmakers or enthusiasts. As an alternative, you may decide to replace the glass with a plastic one purchased from a vendor as previously mentioned.

Early celluloid and plastic dial scales often became brittle and cracked, breaking or chipping over time. Other dial scale damage includes discoloration, burns from pilot lamps and human markings from overzealous DXers. Thankfully there is now a resource for
replacing most any dial. Rock-Sea Enterprises, listed in the resource section of this book, is the only dedicated dial reproduction resource of which the author is aware. They produce an excellent product at a reasonable price. They inventory hundreds of scales and can custom make one they don't stock from what remains of your damaged scale. Rock-Sea also produces the widest variety of reproduction decals available.

## BACKS

Most radios pictured in this book had backs when they were originally sold. Most don't now. The backs were typically flimsy cardboard which became worn and fell off never to be seen again. Often times the antenna was attached to the missing back and is now also missing. To most collectors the absence of a cardboard back will not detract from the value of a radio. Conversely, the presence of a back, particularly if it is in good condition, will add value. One exception to this rule are those radios made during the mid to late 1930s with molded plastic backs (like the Fada 260 series or the Emerson 199). With these radios the back is considered to be a critical element of the radio design and, as such, its absence can devalue a set by as much as 25%.

# PLASTICS DEFINED

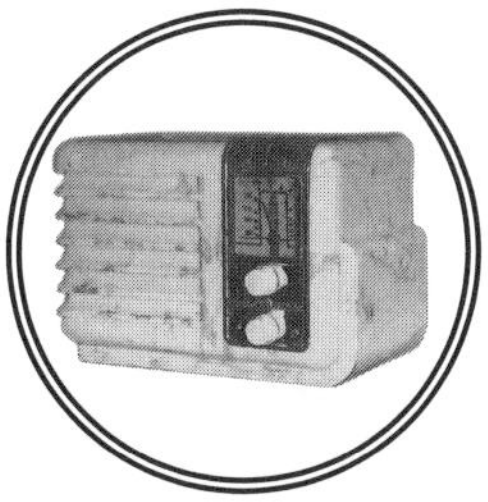

For the novice collector the number of different plastics used in the manufacture of radio cabinets can be confusing. Furthermore, the names we use today have been established more from convention than anything else. That is, most all of the descriptive names currently used were at one time a brand or product name for the material. Typically, many manufacturers produced the same or a similar material. For one reason or another a single brand name has, over time, become synonymous with the material itself. The most common types of plastics and their conventions are as follows:

## BAKELITE

The 'first' plastic, Bakelite was made from a powder and molded under tremendous pressure and heat. It is typically brown or black and somewhat porous. Brown bakelite can be heavily marbleized with various shades of coloration.

## BEETLE

Formed, like bakelite, under heat and pressure, this early plastic can be striking in appearance. Used to mold cabinets from the late 1930s through the early 1940s, beetle is typically opaque ivory in color with marbled streaks of orange, rust, green, blue, red and brown. Some examples may be subtle with just a hint of rust marbling while others evidence themselves in a wide array of deep color tones.

## CATALIN

An early resin, catalin was poured into molds and then cured in low heat ovens over a period of days. Once removed from the mold, if not broken in the process, the catalin radio cabinet was machined to remove debris and to add detail to the design. The labor intensity of this material was not at issue in the late 1930s when it was first introduced in radio cabinetry, but after World War II, with a shortage of manpower, its use quickly became cost prohibitive.

Catalin is most valuable of plastics used in radios. Coloration ranging from opaque solid colors to semi-translucent marbles resembling polished agate. Catalin is prone to chips, cracks, burns (from tube heat), discoloration and shrinking. Such hazards impact the number of catalin sets which have survived and thus price.

## PLASKON

Manufactured using a process similar to bakelite, plaskon was molded in ivory and opaque colors. Typically, radio manufacturers offered the consumer the option of an ivory color at a higher price. Sometimes this was simply a painted bakelite cabinet. Often it was molded plaskon. Occasionally, radios were offered in colors other than ivory. Colors included light and dark greens, red and light blue. Although plaskon wears about as well as bakelite, it is subject to stress lines: superficial cracks.

## UREA

Last of the poured polymer resins. Dominant late 1940s through mid-1950s. Opaque like plaskon typically thicker. Also subject to stress lines, but to a lesser degree.

## PLASTIC

Contemporary injection molded plastics (i.e., polystyrene). First widely used in the late 1940s and available in a wide variety of molded color

# VALUING PLASTIC RADIOS

## CHIPS AND CRACKS

As a rule of thumb, a major flaw in any plastic radio, such as a significant visible chip, crack or warp, will cut the value of that radio in half so long as it remains displayable. Among plaskon radios hairline stress cracks are fairly common. In many models they are the rule and not the exception. As stated previously, so long as the stress lines are not too numerous and do not detract materially from the general aesthetics of the radio, the depreciation should be minimal. On the other hand, the stress free example of a set which is commonly found with stress lines is worth considerably more than average.

## FRAGILITY

Dependent on the thickness of the casting, materials used in construction and the extremity of design, some radios are inherently more delicate than others. The most fragile (such as the Kadette 'Classic') are rarely found in near perfect condition. This 'universe' from which each radio is drawn must be considered in determining its value. In our example, a Kadette 'Classic' in what would appear to the casual observer as marginal condition, might, in fact, be an excellent example of that model, given the condition of other surviving sets.

## REPAIRS

Although there have been some relatively successful attempts at bakelite repair, no repair can go undetected. A well repaired flaw can increase the value of a radio but will never raise its value to that of an unflawed one.

## PAINT CHIPS

While many early plastic radios were available in different colors (typically walnut, black and ivory), oftentimes the ivory colored sets were brown or black bakelite which had been painted at the factory. This factory painting process was similar to that used in the automobile industry. After several layers of paint were applied, the cabinet was baked for several hours. As a result, the paint became 'hard' and much more susceptible to chipping over time. In addition to the problem of chipping, the baking process made the paint that much more difficult to remove. Stripping old baked on paint is extremely difficult and time intensive at best. Be prepared to use caustic chemicals and spend a lot of time. You may then find yourself with a black or brown radio and cast ivory plaskon knobs. This may look fine to you, but it is not 'original' and will be less valuable to other collectors. The net result is that the value of a factory painted radio with paint chips will be reduced by anywhere from ten to fifty percent dependent on the extent and placement of chipping.

# VALUING WOOD RADIOS

Although typically not as popular with collectors in the past, wood cabinet radios have recently gained in value and popularity as the hobby has expanded. It is interesting to note that, when radios were sold in the 1930s and 1940s, it was the wood cabinet which sold for a premium. Plastics were considered then to be a cheap man-made alternative to wood.

**Some wood cabinet basic considerations follow:**

## FINISH

Most wood radios produced during the 1930s through the 1950s were finished with either clear or toned lacquer. The more expensive sets were hand rubbed resulting in the 'piano' finish many collectors today find so desirable. Over the years many things can and usually do happen to a lacquer finish. The finish will dry and chip, peel or separate ('alligator'). Rarely does one come across an original lacquer finish without some evidence of the passage of time. Unless the surviving finish is horrendous, it is worthwhile trying to salvage. There are various products available which will allow you to easily clean the surface and replenish the moisture in the lacquer. Additionally, you may want to amalgamize the finish to cover bare spots. This involves dissolving the original finish with denatured alcohol or other substance and respreading it.

As opposed to clear lacquer, toned lacquers present more of a problem. Many manufacturers, instead of using different wood veneers for contrast, used a single veneer type and sprayed toned lacquers for variations in color and shade. Originally, the difference was difficult to discern. Today it is evident. When toned lacquers age they reveal the original color of the wood underneath, usually in deep contrast. The darker the toner, the more significant the problem. The purchaser of such a radio must resign himself to live with the radio as is or refinish it. Both are a compromise.

## OPTIONAL FINISHES

Until the mid 1930's radio manufacturers assumed the same philosophy as Henry Ford who is noted as saying "You can buy my cars in any color, so long as it's black". They offered no finish options but did often vary cabinets to enhance consumer choice. In 1934 Air King introduced the first crystal (naphthalene) painted wood finish. Beginning in 1935/36 a few manufacturers offered their wood cabinets in optional black lacquer or ivory finishes. These optional finishes were more expensive and were generally custom ordered. They typically came with different trim, often times chrome. The color option never took off with consoles and few were sold. The surviving examples are both striking and significantly more valuable.

The most noteworthy manufacturer offering optional finishes was Zenith. All 1936 models offered with black lacquer. In 1937 Zenith limited the color options to certain models but expanded choice to include ivory. All such radios are extremely rare and highly prized by collectors.

## VENEERS

Most all early radio cabinets were made with inexpensive wood and covered with wood hardwood veneers. Over time the glue which bonds the veneer to the wood of the cabinet can deteriorate. The resulting damage ranges from chipping to separation of large sections. If veneer is separating, but still present careful use of glue and clamps is effective. If pieces of veneer are actually missing, removal and replacement of the remainder of the section is most often the only acceptable cure. If the chip is very small, the hole may be filled with a matching wood putty, shellac stick or toner pencil.

## PAPER VENEER

A concept similar to the toned lacquer, paper veneer involved application of exotic veneer 'decals' to the radio cabinet. This technique was used for both whole radio cabinets and small details. Upon close inspection of the veneer, one can usually discriminate between paper and real veneers. A damaged decal will make the difference apparent. If a radio is ornamented with damaged paper finish, any attempt to repair or refinish the radio will likely remove the decal or damage it further. This should be kept in mind when evaluating such a set for purchase.

## REFINISHING

All but the absolute purists will agree that some radios just need to be refinished in order to be displayable. Many collectors have not seen a well refinished radio and so cannot appreciate the art to it. The process is long and involved, and extremely time intensive, but the result of a 'professional' refinish is astonishing. Of course there is refinishing and there is refinishing. Many collectors will use a caustic solvent to remove the original finish, then stain the cabinet and apply a coat of polyurethane or tung oil. The result is something that looks like it came out of a craft shop. If you are considering the purchase of a radio which is so refinished it should be valued as if it had no finish at all (about half the value of such a set in average condition).

Other types of refinishing are less objectionable and easier to remedy. Oftentimes one encounters a radio with the original finish intact. It is just underneath a coat of slopped on varnish or shellac. Various solvents can be used to remove the top coat and leave the original finish. There is an art to this process in both identifying the correct solvent and its application. Read up on refinishing and talk to other collectors about their experiences before attempting this on a valuable acquisition. The net result on the value of radios so 'refinished' is a reduction by about twenty-five percent.
A note of caution: be certain that the slopped on coat is not polyurethane. If it is, you're back to square one and must remove all finishes.

# COLLECTOR RESOURCES

As the hobby of vintage radio collecting has evolved and expanded over the years so have the resources for collectors. There are regional, national and international clubs. Cottage industries have sprung up offering a wide range of products and services. There are club newsletters, bulletins, references books and monthly periodicals on the subject. And now we have the internet.

## THE INTERNET

With the popularization of the internet, shopping and resourcing for the hobbyist has become both easier and more convenient. With little knowledge and any search engine, one can locate hundreds of radio related websites and find resources for club meetings, supplies, restoration techniques, and other enthusiasts to 'chat' with. A great place to start looking would be one of the websites listed in the following Resource Listings pages of this book. Visit a site and browse the links page to locate other related sites and resources.

In addition to hobby-specific websites, there are now several on-line auction houses, the largest of which, Ebay, has become the primary forum for sales of vintage radios, parts, manuals and related memorabilia. It was only a few years ago that the primary forum for buying and selling vintage radios was paper periodicals, Antique Radio Classified (ARC) being the undisputed leader in the field. Subscribers were at the mercy of the US Mail as to whether they got a shot at a desirable radio for sale. The first to receive and browse through their magazine and make a call to the listing seller bought it. Today, Ebay and the internet has virtually replaced paper classifieds and leveled the playing field for anyone with a computer.

As with the antique business overall, on-line auctions have had a significant impact on the vintage radio hobby. Every collector with access to the internet should spend some time browsing through the auction, dealer and enthusiast sites. Incredible finds turn up on a regular basis and you can search on your own time. The down side is the decreasing number and quality of radios showing up at club meets, antique stores and shows, and even yard and estate sales. Although the days of incredible finds at local yard sales may be over, the savvy collector can still find a bargain with the right tools and knowledge.

## VINTAGE RADIO CLUBS

By far the best part of the hobby is meeting and spending time with other collectors. Club meets provide a forum for exchange of information, purchase of resources and new radio acquisitions. There are few collectors in the United States who are not within a few hours drive from a local or regional club. The following list, although not comprehansive, should help the new collector locate and contact a club nearby.

# VINTAGE RADIO CLUBS

ALABAMA

Alabama Historical Radio Society
PO Box 26452
Birmingham, AL 35226
http://www.bham.net/ahrs/

Southern Vintage Wireless Association
8224 Bailey Cove Rd. SE, Apt. 9
Huntsville, AL 35802

ARKANSAS

Alabama Historical Radio Society
PO Box 26452
Birmingham, AL 35226

Southern Vintage Wireless Association
8224 Bailey Cove Rd. SE, Apt. 9
Huntsville, AL 35802

ARIZONA

Arizona Antique Radio Club
2025 E. LaJolla Dr.
Tempe, AZ 85282-5910
http://www.azarc.org/

Collins Collectors Association
PO Box 10459
Phoenix, AZ 85064

CALIFORNIA

California Historical Radio Society
PO Box 31659
San Francisco, CA 94131
http://www.californiahistoricalradio.com/

CHRS/Sacramento Chapter
PO Box 162612
Sacramento, CA 95816-9998

Military Radio Collector's Group
4141 West L-2
Lancaster, CA 93536

S. California Antique Radio Society
9301 Texhoma Ave.
Northridge, CA 91325-2330

Society of Wireless Pioneers
PO Box 86
Geysrvll, CA 95441

North Valley Chapter CHRS
15853 Ontario Pl.
Redding, CA 96001-9785

Sacramento Historical Radio Society
PO Box 162612
Sacramento, CA 95816-9998

Southern California Antique Radio Society
6934 Orion Ave.
Van Nuys, CA 91406

COLORADO

Colorado Radio Collectors
1985 S. Cape Way
Lakewood, CO 80227

Mountains 'N' Plains Radio Collectors' Association
MPRCA
1249 Solstice Ln.
Fort Collins, CO 80525-1239

CONNECTICUT

Vintage Radio&Communications Museum
33 Mechanic St.
Windsor, CT 06095

Connecticut Area Antique Radio Collectors
500 Tobacco St.
Lebanon, CT 06249

Connecticut Vintage Radio Club
70 Litchfield Rd.
Unionville, CT 06085

FLORIDA

Florida Antique Wireless Group
Box 738
Chuluota, FL 32766

Jacksonville Antique Radio Society
10960 Beach Blvd. Lot 31
Jacksonville, FL 32246

South Florida Antique Radio Collectors
1717 N. Bayshore Dr., Apt. #1549
Miami, FL 33132

GEORGIA

Southeastern Antique Radio Society
PO Box 1055
Tucker, GA 30085-1055
http://www.sarsradio.com/

HAWAII

Hawaii Chapter/AWA
95-2044 Waikalani Pl. C-401
Mililani, HI 96789

Hawaii Historical Radio Club
45 Ala Kimo Dr.
Honolulu, HI 96817-5221

IOWA

Iowa Antique Radio Club
790 Kane St.
Dubuque, IA 52001

ILLINOIS

Antique Radio Club of Illinois
PO Box 1139
La Grange Park, IL 60526
http://www.antique-radios.org/

Belleville Area Antique Radio Club
4 Cresthaven Dr.
Belleville, IL 62221

E. H. Scott Historical Society, Inc.
3712 N. Broadway #4450
Chicago, IL 60613

Illinois Historical Radio Society
8592 N. Creek Rd.
Roscoe, IL 61073

Museum of Broadcast Communications
78 E. Washington
Chicago, IL 60602

E.H. Scott Historical Society Inc.
3712 N. Broadway #4450
Chicago, IL 60613

INDIANA

Indiana Historical Radio Society
245 N. Oakland Ave.
Indianapolis, IN 46201-3360

KANSAS

Mid-America Antique Radio Club
12309 W. 70th Terr.
Shawnee, KS 66216

Xtal Set Society
789 N. 1500 Rd.
Lawrence, KS 66049-9194

KENTUCKY

Cumberland Valley Radio Society
HC-66 Box 1604
Barbourville, KY 40906

Kentucky Chapter/AWA
3114 Boxhill Ct.
Louisville, KY 40222

LOUISIANA

Louisiana&Gulf Coast Antique Radio Club
750 Moore St.
Baton Rouge, LA 70806

MASSACHUSETTS

Greater Boston Antique Radio Collectors
PO Box 2-W20
Carlisle, MA 01741

New England Antique Radio Club
PO Box 918
Melrose, MA 02176
http://www.nearc.net/MARYLAND

Radio History Society, Inc. Museum and library
1205 Gladstone Dr.
Rockville, MD 20851

MICHIGAN

Michigan Antique Radio Club
3520 Okemos Rd., #6 Suite 100
Okemos, MI 48864-5943
http://michiganantiqueradio.org/home.jsp

MINNESOTA

Northland Antique Radio Club
PO Box 18362
Minneapolis, MN 55418
http://www.geocities.com/TelevisionCity/4544/

MISSOURI

Antique Radio Collectors & Historians
509 Main Street
Fenton, MO 63049

Four States Antique Radio Club
1702 S. McKinley Ave.
Joplin, MO 64801

MISSISSIPPI

Mississippi Historical Radio and Broadcasting Society
2412 C Street
Meridian, MS 39301

NORTH CAROLINA

Carolinas Chapter/AWA
PO Box 3015
Matthews, NC 28106
http://www.cc-awa.org/

East Carolina Antique Radio Club
218 Bent Creek Dr.
Greenville, NC 27834

Vintage Radio Unique Society
312 Auburndale St.
Winston-Salem, NC 27104

NEBRASKA

Nebraska Antique Radio Collectors Club
905 West First
North Platte, NE 69101

Midwest Radio Club MRC
PO Box 6291
Lincoln, NE 68506-0291

NEW HAMPSHIRE

Military Radio Collectors Association
PO Box 467
Holderness, NH 03245

NEW JERSEY

International Antique Radio Club
PO Box 5367
Old Bridge, NJ 08857

New Jersey Antique Radio Club
13 Cornell Pl.
Manalapan, NJ 07726
http://www.njarc.org/

NEVADA

Nevadan Enthusiasts for Antique Radios
1881 Capo San Vito Ave.
Las Vegas, NV 89123

NEW MEXICO

New Mexico Radio Collectors Club
39 Chaco Loop
Sandia Park, NM 87047

NEW YORK

Antique Wireless Association
Box E
Breesport, NY 14816
http://www.antiquewireless.org/

Greater NY Vintage Wireless Association
52 Uranus Rd.
Rocky Point, NY 11778

Hudson Valley Antique Radio and Phonograph Society-AWA
PO Box 207
Campbell Hall, NY 10916

Niagara Frontier Wireless Association
135 Autumnwood
Cheektowaga, NY 14227

Schenectady Antique Radio Club
Bldg. 20, Apt. 1
Troy, NY 12182
http://dadellers.tripod.com/sarc/
(Beware of Pop-Ups!)

Tri-State Antique Radio Club
Box 172
Valley Cottage, NY 10989

Central New York/Northern Pa.Antique Radio Club
711 Elm St.
Groton, NY 13073

OHIO

Antique Radio Collectors of Ohio
2929 Hazelwood Ave.
Dayton, OH 45419

Buckeye Antique Radio and Phonograph Club
1716 Chestnut Blvd.
Cuyahoga Falls, OH 44223

Central Ohio Antique Radio Association
3782 Mill Stream Dr.
Hilliard, OH 43026

Cincinnati Antique Radio Society
6 Indian Dr.
Sardinia, OH 45171

Society for the Preservation of Antique Radio Knowledge
PO Box 292111
Kettering, OH 45429

Cincinnati Antique Radio Collectors
6805 Palmetto
Cincinnati, OH 45227

OKLAHOMA

OKVRC Green Country Chapter
5830 E. 58th St.
Tulsa, OK 74135-8131
http://members.cox.net/okvrc/

Oklahoma Vintage Radio Collectors Club
PO Box 50625
Midwest City, OK 73140-5625
http://members.cox.net/okvrc/

OREGON

Northwest Vintage Radio Society
PO Box 82379
Portland, OR 97282-0379

PENNSYLVANIA

Central Pa. Radio Collectors Club
1045 Bonair Dr.
Williamsport, PA 17701

Collins Radio Association
PO Box 34
Sidman, PA 15955

Delaware Valley Historic Radio Club
PO Box 5053
New Britain, PA 18901
http://pw1.netcom.com/~firstake/dvhrc.htm

Pittsburgh Antique Radio Society, Inc.
Ruthfred Dr.
Pittsburgh, PA 15241

RHODE ISLAND

Rhode Island Antique Radio Enthusiasts
61 Columbus Ave.
N. Providence, RI 02911

TENNESSEE

Memphis Antique Radio Club
459 Second St.
Henderson, TN 38340
http://marcradioclub.tripod.com/MARCHome.htm

Mid-South Antique Radio Collectors
933 Venice Rd.
Knoxville, TN 37923

Nashville Vintage Radio Club
5026 Suter Dr.
Nashville, TN 37211

Music City Vintage Radio & Phonograph Society
MCVR&PS
PO Box 22291
Nashville, TN 37202

TEXAS

Houston Vintage Radio Association
PO Box 31276
Houston, TX 77231-1276

Texas Antique Radio Club
218 Shannon Lee
San Antonio, TX 78216
http://www.gvtc.com/~edengel/TARC.htm

Texas Panhandle Vintage Radio Society
4086 Business Park Dr.
Amarillo, TX 79110

Vintage Radio & Phonograph Society
PO Box 165345
Irving, TX 75016

VERMONT

Border Amateur Radio Club
PO Box 372
Derby, VT 05829

WASHINGTON

Puget Sound Antique Radio Association
PO Box 2095
Snohomish, WA 98291-0125

Radio Enthusiasts of Puget Sound
14911 Linden N.
Seattle, WA 98133

Vintage Audio Listeners and Valve Enthusiasts
1127 N.W. Bright Star Ln.
Poulsbo, WA 98370

WEST VIRGINIA

Museum of Radio & Technology
1640 Florence Ave.
Huntington, WV 25701

W. Va. Chapter, AWA
405 8th Ave.
St. Albans, WV 25177

WISCONSIN

Wisconsin Antique Radio Club
10230 W. Greenwood Ter.
Milwaukee, WI 53224-3716

Western Wisconsin Antique Radio Collectors Club
1611 Redfield St.
La Crosse, WI 54601

***Club listings provided courtesy of RadioSwapMeet.com**, a multi-faceted website for vintage radio enthusiasts. The site includes Radio Clubs and Oranization; Forum area for discussions; Photo Gallery; Electronic Restoration Tips; Articles; Vacuum Tube Exchange; schematics and more.*

# RESOURCE LISTINGS

## GENERAL RESOURCES

**All Things Radio** *www.allthingsradio.com*
Over 2600 products including vintage and repro radios, restoration supplies, equipment, tools and professional services, research, old time broadcasts on CD, new and vintage books, chat room, newsletter......more!

**Radiomania.com** *www.radiomania.com*
Publisher of the leading vintage radio identification and price guides. Books available signed by the author and shipped without charge. On-line appraisal service. Check website for details. *rpbook@bellatlantic.net*

**CollectingChannel.com**
Web site sponsored by *Maloney's Antiques & Collectibles Resource Directory*. Provides collecting related articles, web casts, television and audio programming, and a Collector's Directory with over 600,000 resources. 808 Live Oak Dr., Ste. 126, Chesapeake, VA 23320
Phone: 757-321-1450 Fax: 757-321-1451

**RadioSwapMeet.com**
Multi-faceted website for vintage radio enthusiasts. Includes Radio Clubs and Oranization; Forum area for discussions; Photo Gallery; Electronic Restoration Tips; Articles; Vacuum Tube Exchange; schematics and more!

## PRODUCTS / SUPPLIES

**TheOldRadioFixerUpperGuy** *www.oldradiofixerupperguy.com*
Dedicated to the repair of both wooden and Bakelite radios from days gone by. We perform both electronic and cosmetic restorations, and specialize in the repair of shattered Bakelite.

**Antique Electronic Supply** *www.tubesandmore.com*
Tubes, books, transformers, sockets, capacitors, resistors, literature, cabinet restoration materials, wire, grill cloth, tools, information, much more. *info@tubesandmore.com*
6221 S Maple Avenue, Tempe, AZ 85283 Phone: 480-820-5411

**Constantine's** *www.constantines.com*
Complete line of woodworking and refinishing tools, parts and supplies encompassing all refinishing needs.
*customerservice@constantines.com*
2050 Eastchester Rd., Bronx, NY 10461 Phone: 800-223-8087

**Rock-Sea Enterprises** *members.aol.com/RockSeaEnt*
Antique radio dial scales & decals, all types. Catalog inventory and custom made. Quarterly newsletter. *dials@juno.com*
PMB 241, 323 -110 E. Matilija St., Ojai, CA 93023

**Classic Radio Gallery** *cpu.net/classicradio*
Schematic services available for your antique radios. On-line antique radio museum of some of the high-points in radio design. Thumbnail photos in various catagories with links to screen-size photos.

**Old Time Replications**
Wide spectrum of reproduction knobs, pushbuttons, escutcheons, grills and other cast parts. *oldtimerep@aol.com*
5744 Tobias Ave., Van Nuys, CA 91411 Phone: 818-786-2500

**Curtis A. Lutz**
Rubber U-channel for Zenith & other dial glasses, $1.25/ft, postpaid, 30 ft or more only $1.00/ft. Also new transformers for building DC power supplies. *ANTIQRADIO@aol.com*
101 N. Hickory Place, Broken Arrow, OK 74012. Phone: 918-251-4915

**Antique Radio Grille Cloth HQ** *www.grillecloth.com*
The number one source for reproduction grille cloth for antique radios from 1920 thru 1940. *john@grillecloth.com*
624 Cedar Hill Road, Ambler, PA 19002

**Doyle Roberts**
Clear plastic dial lense reproduction from damaged old lens or drawing of dial opening. *no e-mail address available*
HC63, Box 236-1, Clinton, AR 72031

## DEALERS / RADIOS FOR SALE

**All Things Radio** *www.allthingsradio.com*
Over 2600 products including vintage and repro radios, restoration supplies, equipment, tools and professional services, research, old time broadcasts on CD, new and vintage books, chat room, newsletter......more!

**CatalinRadio.com**
CatalinRadio.com offers a museum, interesting information and an impressive sales gallery that offers the largest quality selection of catalin and art deco radios anywhere in the world. Parts and services are also available on this website.

**RadioSalon.com**
Mid-century modern electric radios and phonographs, fine plastics collectibles. Focus on plastic items produced between 1937 to 1960. All items are refurbished electronically and/or cosmetically.

**EuropRadio.com** *www.europradio.nl*
Sponsored by collector in the Netherlands. Vintage European radios for sale. Tabletop, portable and car radios. All radios are in perfect playing condition. Prices include worldwide shipping.

**RadioAttic.com**
Almost 300 radios & radio-related items for sale. Sell your radio at the Radio Attic; your ad stays on-line until your radio sells!

**Radiomania.com**
On-line showroom specializing in sales of rare and unusual wood art deco table radios, clocks, lighting, Chase and more industrial design icons. Prop rentals. *rpbook@bellatlantic.net*

**NewEraAntiques.com**
On-line showroom with art deco focus. Radios, televisions, telephones, lighting, Chase. Brooklyn, NY.

**Great Northern** *GN4Radios.com*
Specializing in Zenith and EH Scott. Also restoration parts: original and reproduction. *Alan@gn4radios.com*
5200 Bloomington Ave., Mnpls, MN 55417 Phone: 612-727-2489

**Urban Radios & Collectibles** *www.urban-antiqueradio.com*
Sales of quality antique radios that work. Backed by professional testing, restoration and warranty. Vintage Hi-Fi included. Located in Westport, CT

**Alaska Antique Radios** *www.AlaskaAntiqueRadios.com*
Internet sales of antiques, specializing in art deco design antique radios from the 1930's, 1940's, and 1950's.

**Aborn True Value** *www.aborntruevalue.com/aborn/default.asp*
Specializing in antique radios 1930 - 1960 and collectible jukeboxes 1938 - 1961. Exclusive Wurlitzer Jukebox distributor for new jukeboxes in MA. *aborntv@earthlink.net* Jerry Katz
438 Harvard St. Brookline, MA 02446 Phone: 617-739-9000

**John Kendall's Vintage Electronics** *vintage-electronics.com*
Wholesaler of restorable collectible radio, audio, and other vintage electronic items to the collector/restorer, and to dealers, internationally since 1980. Also related books, schematics, tubes, used original parts.
*JohnMS@vintage-electronics.com* Baltimore,MD

**Primeau Music** *primeaumusic.com*
Specializing in rare and collectible bakelite tabletop radios and record players, reconditioned and painted in colors of the era. Focus on Deco and Machine Age. *Sales@primeaumusic.com*
916 Oxford Street West; London, Ontario, Canada N6H 1V3
Phone: 519-657-4433 Fax: 519-471-4095

**Radio Age** *radioage.com*
Internet sales of antique radios, specializing in radios with the Art Deco design style of the 1930s-1940s. *info@radioage.com*
Terry Williams, 2751 Caminito San Pablo, Del Mar, CA 92014
Phone: (888) 339-4517 Fax: (603) 843-7903

## DEALERS/RADIOS FOR SALE (CONTINUED)

**Radio Age** *radioage.com/products0.html*
Internet sales of antique radios, specializing in radios with the Art Deco design style of the 1930s 1940s.

**AMradios.com** *www.amradios.com*
Old tube radios & schematic service. Radios available from the 1920's - 1950's plus much more. *Charles@AMradios.com*
Located in South Florida 954-434-7758

**International Vintage Wireless Exchange**
Ability to source and supply radios from every genre and every country. Can help any interested collector to buy, sell or trade. Enormous stock, enormous database. Radios are my passion. *RADIOZE@aol.com*
Finchcroft, Broadwater Down, Royal Tunbridge Wells, Kent TN2 5PE, United Kingdom.
Phone: ++44.1892.543505 Fax: ++44.1892.543505

**RogRadCo** *crystalrad.com*
Specialized in sales of crystal radios, kits, parts and accessories. Also knowledgeable in dates of production of antique crystal radios. *rogradco@excite.com* P.O. Box 813, Dearborn Heights, MI 48127. Phone: (313) 277-2529

**All Things Radio** *www.allthingsradio.com*
Over 2600 products including vintage and repro radios, restoration supplies, equipment, tools and professional services, research, old time broadcasts on CD, new and vintage books, chat room, newsletter......more!

## RESTORATION SERVICES

**NED Service/Dan Collins** *www.radiorestore.com*
Repair/Restoration Services - Large NOS parts inventory. Plastic/wood/metal parts reproduction/fabrication. Specialize in 1 930-1945 radios. *col@bellatlantic.net* P.O.Box#1026, Southampton, PA 18966. Phone: (215) 364-9644

**Great Northern** *GN4Radios.com*
Complete guaranteed radio restoration, including speaker restoration, specializing in 1930s and 1940s. Supplier of restoration parts, original & reproduction. *Alan@gn4radios.com*
5200 Bloomington Ave., Minneapolis, MN 55417
Phone: 612-727-2489 Fax: 612-821-4821

**Ken's Antique Electronics**
Restore and collect radios from the 1920's to the 1960's. Specialize in Zenith 1930's and 40's. Cabinet and chassis restoration on all radios. Buy & sell all types. *kszenith@webtv.net*
871 River Road, Eugene, Oregon 97404. Phone: 541-485-8676

**Antique Radio Restoration & Repair** *www.oldradiodoc.com*
Professional restorations for all tube type antique table radios, communication receivers and amplifiers. Complete overhauls. Cabinet refinishing. *radiodoc@neca.com* 20 Gary School Road, Pomfret Center, CT 06259 Phone: 860-928-2628

## EXPERTS / SPECIALISTS

**Radiomania Appraisal Service** *www.radiomania.com*
Expert valuation and certified appraisals for radios, clocks, and decorative pieces categorized as 20th century modern. On-line, mail or personal service. Competitive rates. *rpbook@bellatlantic.net*
2109 Carterdale Road, Baltimore, MD 21209

**Classic Radio Gallery** *cpu.net/classicradio*
On-line antique radio museum of some of the high-points in radio design. Thumbnail photos in various catagories with links to screen-size photos. Schematic services available for your antique radios. *mmabbs@cpu.net*

**Don's Antique Radio** *hometown.aol.com/dmradios*
Don's home page with links to tube and transistor radios for sale. Radio Miscellanea for sale. Radio Museum. *Dmradios@aol.com*
D. Maurer, 29 South 4th Street, Lebanon, PA 17042

## COLLECTIONS AND MUSEUMS

**Edwin G. Cook**
Collector of table model radios of the mid and late 1930s and early 1940s specializing in Ingraham cabinet of radios of all known makes. *hopopco@aol.com*
3215 NE 111th Circle, Vancouver, WA., 98686. Phone: 360-573-1439

**Francisco Haro**
Restore and collect bakelite and tabletop radios from 1930's to mid 1950's, primarily American, some European. Mexico, D.F. *fcoharo@yahoo.com*

**Maurice Giroux/Radio 1920**
Serious "1920's radio collector" from Montreal. Favorites are Atwater Kent and Amplion horns. Have large collection, but always looking for new items. Often trade plastic radios (addison, catalins) for wants. *giroux1@total.net*
559 Hudon, Laval, PQ, Canada H7P2L4. Phone: 450-963-1601

**Michael Adkison/Wet Shoe Radios**
Collect interesting, rare, strange or just beautiful radios, primarily 1935-1942. Minor repairs, restoration and information. Always looking for old parts, tubes and orphan radios; neglected or not. *adkison@home.com*
3366 Ken Lake Drive SW, Olympia, WA 98512-5522

**Old Radio Zone** *www.oldradiozone.com*
Large private online collection of 1930s consoles, cathedrals and tombstones with an emphasis on high-end Zenith's. See Walton's, Robotdials and Stratospheres! Always buying quality radios, parts and memorabilia. *oldradiozone@aol.com*

**Jeff Conley**
Collector and restorer of vintage radios, specializing in quality wood radios of the 1930's and 40's. Toledo, Ohio. *jconley@accesstoledo.com* Phone: (419) 838-5866

**noveltyradio.com** *www.noveltyradio.com*
Gary Arnold's transistor novelty radios for sale. Accurate listings. Click to view the picture of any radio. *gary@noveltyradio.com*
615 oak street Marion NC 28752 Phone: 828-652-6893

**Bruce Eddy**
Collecting with interests in unusual 1930's particularly globe dials, high tube counts, motor tuning, and in 1950's variable selectivity radios. *eddyb@voyager.net*
2521 W. Needmore Hwy., Charlotte, MI 48813

**Ignacio Arturo Diez**
Collecting Old radios,reel to reel tape recorders, reel tapes, 8 , super 8 and 16 mm. movies, projectors and cameras. Old photographic cameras & complements. Also all the "ZENITH TRANSOCEANIC" Series. *olivaresp@netcombbs.com.ar*
PO Box 187, Rio Grande, Tierra Del Fuego 9420, Argentina
Phone: 54- 2964- 15 564 400 Fax: 54- 2964- 42 1502

**Peter J. Malvasi, W2PM**
*members.aol.com/pmalvasi/keypix/pete.html*
Wireless and Telegraph History and Collection Information
*pmalvasi@aol.com*

**Vic Johnston** *members.xoom.com/antiqun/Vic'sOR.html*
Personal collection of radios 1914-1971, consists of wireless, battery, Scotts, radio and TV games, toys and related items. Amazing Stories magazines, aviation and military items and Tvs. *antiqun@gateway.net*

**Art's Antique Radios** *www.ntsource.com/~oldradio99*
Collection specializing in the unique and any Coke or Pepsi radio. Restoration services also available at reasonable rates. Will buy single radios or whole collections. Chicago, Illinois. Phone: 630-739-1060

**Antique Electronics** *www.geocities.com/antiquelectronics*
David S. Smith, collector of wireless era through battery sets of the 1920's. Also sells and trades items from this era. Interested in parts for building breadboard sets and manuals from this period. *dblas@cgate.net*

**Radiolaguy.com** *www.radiolaguy.com*
Virtual museum provides information concerning vintage radio & television. Also offers radios, televisions and related items for sale. Washington State. *radios@radiolaguy.com* Phone: 360- 834-5741

## COLLECTORS (CONTINUED)

**RadioRich**
Collector-dealer for 10 years in the NY/NJ/CT tri-state area. Buy-sell-lease, tube & transistor set radios, plus parts, radio manufacturers' signs & advertisements. *radiorich@prodigy.net*
154 Hudson Terrace, Piermont, New York, 10968.
Phone: 914-359-3809

**TRC Electric Antiques** *trcelectricantiques.hypermart.net*
Mid-century modern electric radios and phonographs, fine plastics collectibles. Focus on plastic items produced between 1937 to 1960. All items are refurbished electronically and/or cosmetically.

**Greg Farmer** *geocities.com/SiliconValley/3839/radio*
The Farmer's Old Radio Notebook highlights some unusual items including Midwest and International Kadette, early battery radios, unusual radio speakers, and bicycle radios. *gif@usfamily.net*

**Airwaves Wayne's Vintage Radios** *execulink.com/~michiels*
Radios bought, sold and traded on-line. View Canadian radios like Addison, Sparton,Rogers and more. *michiels@oxford.net*
Wayne Michiels, 7 Lowrie Cres, Tillsonburg, Ontario, Canada, N4G 5L4 Phone: 519- 688-0964

**oldradios.com** *www.oldradios.com*
George Kaczowka's collection of US radios of the 1930s, transistor radios of the 1950s, international Table radios of the 1950s.
*gsk@oldradios.com* P.O. Box 103, Boylston, MA 01505
Phone: (508) 869-6376

**Frank Arno** *home.att.net/~arnocasa/wsb/index.html*
Radio collection consisting primarily of table top's & portables along with "other collectibles". New inventory is periodically added.
Garden City, NY *arnocasa@worldnet.att.net*

**Skywaves** *www.connix.com/~harry/radio.htm*
Website dedicated to use of vintage radios for AM and SW listening. A place for radio operators looking to enhance the performance \of their vintage sets. *skywaves@harry.connix.com*
Harry White, POB 137, Colebrook, CT 06021

**Simon Wade**
Comprehensive collection of classic design radios from all corners of the world. Always keen to trade pre-broadcast, 1920s, 1930s, colour, shape and style. Always willing to help and advise.
*Simonwade@aol.com* Finchcroft, Broadwater Down, Royal Tunbridge Wells, Kent TN2 5PE, United Kingdom.
Phone: ++44.1892.543505 Fax: ++44.1892.543505

**Radiotiques** *nealsnet.pair.com*
Web site dedicated to the preservation and restoration of antique tube radios. *Jeffrey.Goldstein@astrazeneca.com*

**Frank's Radio Page** *home.att.net/~arnocasa/wsb/html/view.cgi-home.html-.html*
Radio collection consisting primarily of table top's & portables along with "other collectibles". New inventory is periodically added.

**South Florida Antique Radio Collectors**
Organization dedicated to the preservation of antique radios. Members from all over Florida, mainly Miami/Ft. Lauderdale/Boca Raton area. Quarterly swap meets. Publish quarterly newsletters and a restoration Resource Guide. c/o Tom Valenti, President, 1717 North Bayshore Drive #1549, Maimi, FL 33132
*bmenassa@worldnet.att.net, valenti9@bellsouth.net*

**RL Drake Museum** *dproducts.be/drake_museum*
On-line Museum dedicated to RL DRAKE Radio Telecommunications Equipment (est. 1943). Includes detailed history of company, products and technology. Excellent resource. *jmc@dproducts.be*

## CRYSTAL RADIOS

**RogRadCo.** *crystalrad.com*
Specialized in sales of crystal radios, crystal radios kits, crystal radios parts, and accessories. Also knowledgeable in dates of production of antique crystal radios. *rogradco@excite.com*
P.O. Box 813, Dearborn Heights, MI 48127.
Phone: (313) 277-2529

If you would like to share information about your radio-related business or collecting interests in this resource forum of future Radiomania books, please contact us for details.

**Radiomania® Books**
**2109 Carterdale Road**
**Baltimore, MD 21209 U.S.A.**

E-mail: rpbook@bellatlantic.net
Fax: 800-891-4484

# RADIO LISTINGS

The following pages comprise the identification and price guide section of this reference. All listings are arranged alphabetically by manufacturer and then numerically by model number within each manufacturer.

---

Each radio is pictured in one of its variations with description including the following information:

Manufacturer Brand Name
Model Number(s) and/or Model Name(s)
Year of Manufacture (c.)
Description of Cabinet Material(s)
Model Variations (if any)
Tube Count
Band Count
Special Features/Options
Current Market Value(s) Including Variations

In addition to the above, this book introduces easily recognizable symbols to indicate specific important attributes of a radio:

Manufactured in Canada: Canadian radios appear frequently within the United States and were often manufactured by US companies for the Canadian market.

Industrial Designer is known or attributed. The name of the designer is specified with an indication as to whether the designer is documented (Des.) or attributed (Attrib.)

Wood cabinet made by E.Ingraham Co., Bristol, CT. Such radios are considered collectible in themselves. Ingraham's level of design and craftsmanship, Their ability to contour wood panels was unequaled. Ingraham cabinet radios are among the highest valued.

Manufactured to commemorate the 1933 Chicago World's Fair, 'A Century of Progress'(C.O.P.). Also denotes radios sold at the world's fair but not designed specifically as a commemorative.

Manufactured to commemorate the 1939 New York World's Fair, 'The World of Tomorrow'. Also denotes radios sold at the world's fair but not designed specifically as a commemorative.

---

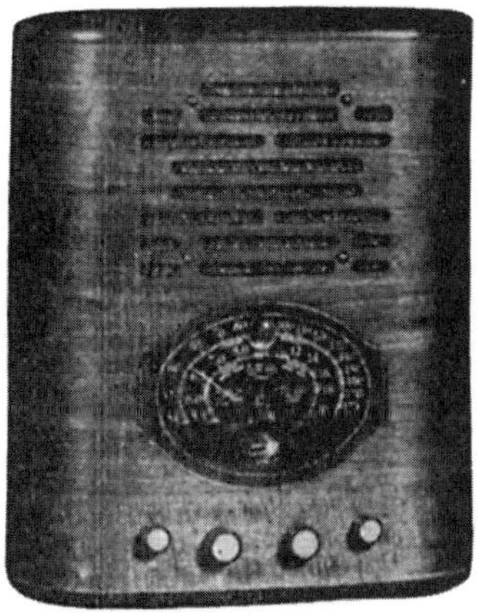

ACE

**7 DELUXE, c.1936**
7 tubes, 2 bands
$325

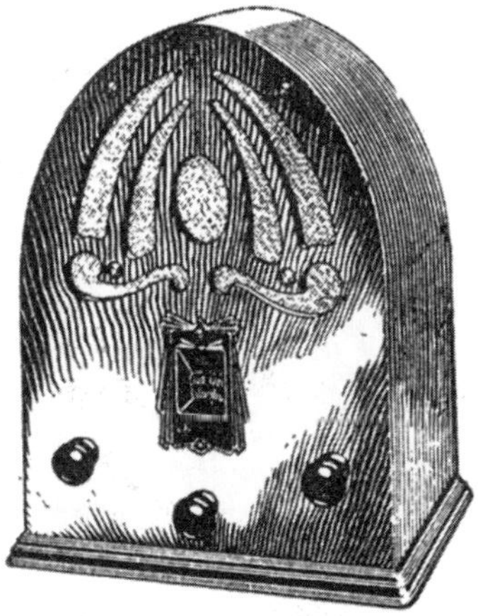

ACE

**MULTI-MU, c.1932**
5 tubes, 1 band
$200

ACE

**5, c.31**
6 tubes, 1 band
$175

ACE

**7 LOWBOY, c.1936**
7 tubes, 2 bands
$175

INGRAHAM INGRAHAM

ACE

**51G / 51D, c.1936**
5 tubes, 1 band
51G $175 / 51D $325

Acratone

**5B 'DANDY', c.1936**
5 tubes, 1 band
$75

Acratone

**5D, c.1936**
5 tubes, 1 band
$125

Acratone

**6B, c.1936**
5 tubes, 2 bands
$110

Acratone

**7B, c.1936**
5 tubes, 2 bands
$150

Acratone

**7C, c.1936**
5 tubes, 3 bands
$150

Acratone

**25B 'BUDDY', c.1936**
4 tubes, 1 band
$60

Acratone

**25D, c.1936**
4 tubes, 1 band
$65

Acratone

**34C, c.1936**
6 tubes, 3 bands
$125

Acratone

**40D, c.1936**
6 tubes, 1 band
$225

Acratone

**53, c.1934**
Repwood
4 tubes, 1 band
$425

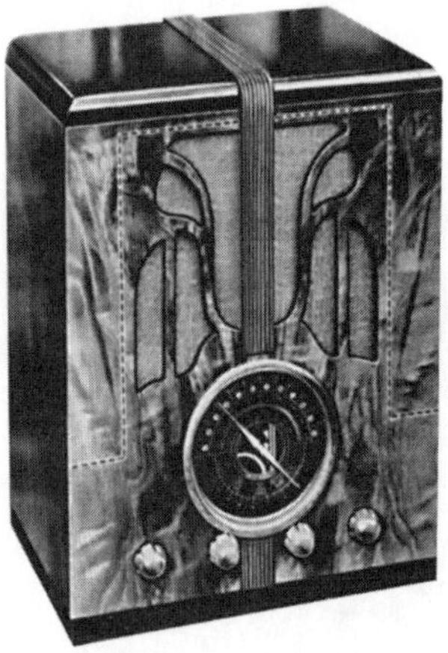

Acratone

**53D, c.1936**
6 tubes, 3 bands, DC
$75

Acratone

**56, c.1934**
4 tubes, 1 band
$50

Acratone

**56D 'LOBOY', c.1936**
6 tubes, 3 bands
$75

Acratone

**58, c.1934**
4 tubes, 1 band
$115

Acratone

**60, c.1934**
4 tubes, 2 bands
$325

Acratone

**63, c.1934**
5 tubes, 1 band
$100

Acratone

**65, c.1934**
5 tubes, 1 band
$450

Acratone

**67, c.1934**
5 tubes, 1 band
$135

Acratone

**75, c.1934**
5 tubes, 2 bands
$150

Acratone

**77, c.1934**
6 tubes, 2 bands
$250

Acratone

**93/96, c.1934**
7 tubes, 2 bands(93), 3 bands (96)
$175

Acratone

**117, c.1934**
6 tubes, 4 bands
$350

Acratone

**118, c.1934**
6 tubes, 1 band, DC
$125

Acratone

**168B, c.1936**
5 tubes, 2 bands, DC
$65

Acratone

**189K, c.1938**
4 tubes, 1 band, AC/DC
$50

Acratone

**191K, c.1938**
6 tubes, 2 bands, radio/phono
$125

Acratone

**192K, c.1938**
6 tubes, 2 bands, AC/DC
$75

Acratone

**194-6K, c.1938**
4 tubes, 1 band, AC/DC
$350

Acratone

**197K, c.1938**
5 tubes, 1 band, AC/DC
$75

Acratone

**198K, c.1938**
6 tubes, 2 bands, AC/DC
$100

Acratone
**201/203K, c.1938**
5 tubes, 1 band, AC/DC
walnut bakelite $175
ivory plaskon $250

Acratone
**204K, c.1938**
6 tubes, 1 band, DC
$30

Acratone
**206K 'World's Fair', c.1939**
6 tubes, 2 bands
$150

Acratone
**207K 'World's Fair', c.1939**
8 tubes, 3 bands
$225

Acratone
**209K, c.1938**
7 tubes, 2 bands, AC/DC
$75

Acratone
**211K, c.1938**
8 tubes, 3 bands
$300

Acratone
**213K, c.1938**
5 tubes, 2 bands, DC
$75

Acratone
**214K, 1938**
7 tubes, 2 bands, AC/DC
$100

Acratone
**216K, c.1938**
6 tubes, 2 bands, AC/DC
$75

Acratone

**217K, c.1938**
7 tubes, 3 bands
$75

Addison

**55, c.1949**
5 tubes, 1 band
$50

Addison

**61, c.1953**
4 tubes, 1 band
$40

Addison

**63, c.1953**
5 tubes, 1 band
$60

Addison

**64, c.1953**
6 tubes, 1 band
$125

Admiral

**6R10,11, c.1949**
6 tubes, AM/FM, Radio/Phono
6R10 black bakelite $75
6R11 brown bakelite $60

Admiral

**7RT41-6L1, c.1947**
6 tubes, 1 band, radio/phono
bakelite $75

Admiral

**7RT42-6L1, c.1947**
6 tubes, 1 band, radio/phono
$50

Admiral

**7RT43-6L1, c.1947**
6 tubes, 1 band, radio/phono
mahogany $60
walnut $50

Admiral

**7T04-5K1, c.1947**
5 tubes, 1 band
$25

Admiral

**7T06-4B1, c.1947**
4 tubes, 1 band, DC
$20

Admiral

**7T10-CE/EC, c.1947**
5 tubes, 1 band
black bakelite $60
painted ivory $50

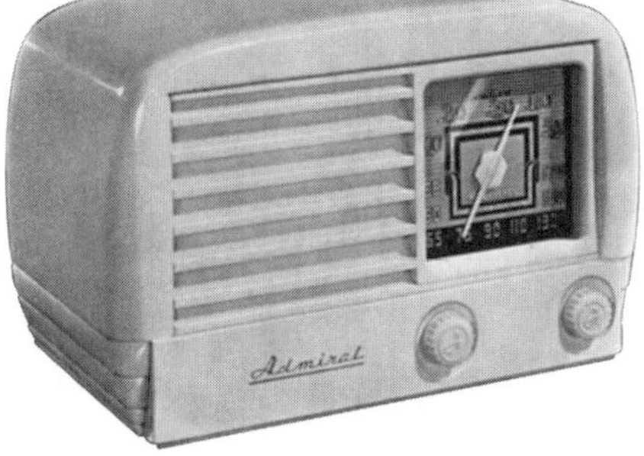

Admiral

**7T10-5K1, c.1947**
5 tubes, 1 band
brown bakelite $40
black bakelite $50
painted ivory $50

Admiral

**7T10-5N1, c.1947**
5 tubes, 1 band
brown bakelite $35
black bakelite $45
painted ivory $40

Admiral

**7T12-4B1, c.1947**
4 tubes, 1 band, DC
brown bakelite $25

Admiral

**10-A5, c.1940**
5 tubes, 2 bands, DC
$25

Admiral

**12-B5,14-B5, c.1941**
5 tubes, 1 band
12B-5 brown bakelite $75
14B-5 ivory plaskon $125

Admiral

**15-D5, c.1941**
5 tubes, 1 band
brown bakelite $125

Admiral

**23-A6, c.1940**
6 tubes, 1 band, AC/DC
$60

Admiral

**40-A4, c.1940**
4 tubes, 1 band, DC
$20

Admiral

**41-B4, c.1940**
4 tubes, 1 band, DC
$25

Admiral

**47-J55, c.1941**
5 tubes, 2 bands
$50

Admiral

**48-J6,49-J6, c.1941**
5 tubes, 2 bands
48-J6 brown bakelite $45
49-J6 painted ivory $50

Admiral

**50-J6, c.1941**
5 tubes, 2 bands
$90

Admiral

**51-J55,52-J55, c.1941**
5 tubes, 2 bands
51-J55 brown bakelite $40
52-J55 painted ivory $40

Admiral

**51-K6,52-K6, c.1941**
5 tubes, 2 bands
51-K6 brown bakelite $40
52-K6 painted ivory $40

Admiral

**53-K6, c.1941**
6 tubes, 2 bands
$90

Admiral

**54-XJ55, c.1941**
5 tubes, 1 band, radio/phono
$50

Admiral

**55-A7, c.1941**
7 tubes, 2 bands
$45

Admiral

**64-K5, c.1941**
5 tubes, 1 band, radio/phono
$150

Admiral

**69-M5, c.1941**
5 tubes, 1 band, radio/phono
$40

Admiral

**70-N6, c.1941**
6 tubes, 1 band, radio/phono
$150

Admiral

**113,4,5-5A, c.1938**
5 tubes, 1 band, AC/DC
113-5A black bakelite $175
114-5A brown bakelite $150
115-5A ivory plaskon $250

Admiral

**123,4,5,6-5E, c.1938**
5 tubes, 1 band
123-5E black bakelite
124-5E brown bakelite
125-5E ivory plaskon
126-5E red plaskon

Admiral

**141-4A, c.1939**
4 tubes, 1 band, DC
$20

Admiral

**153-5L, c.1940**
5 tubes, 1 band, radio/phono
$75

Admiral

**384-5S, c.1940**
5 tubes, 1 band
$75

Admiral

**389-5S, c.1940**
5 tubes, 1 band
$60

Admiral

**516-5G/I, c.1938**
5 tubes, 1 band
516-5G brown bakelite $60
516-5I ivory plaskon $90

Admiral

**520-5C, c.1939**
5 tubes, 1 band, radio/phono
$65

Admiral

**521-5C, c.1938**
5 tubes, 1 band, radio/phono
$60

Admiral

**985-5Z, 990-5Z, 985-6Y, 990-6Y, c.1938**
5/6 tubes, 1 band
985-5Z, 985-6Y black bakelite $200
990-5Z, 990-6Y ivory plaskon $275

Admiral

**AW11, c.1939**
5 tubes, phono remote
$65

Admiral

**L767, c.1936**
7 tubes, 3 bands, DC
$75

Admiral

**M169, c.1936**
6 tubes, 3 bands, DC
$90

Admiral

**X641, c.1936**
6 tubes, 3 bands, DC
$75

AERO

**HI-MU, c.1932**
5 tubes, 1 band
$150

AIR CASTLE

**264, c.1947**
4 tubes, 1 band, DC
$20

AIR CASTLE

**281, c.1947**
6 tubes, 2 bands
$30

AIR CASTLE

**287,328, c.1947**
5 tubes, 1 band
287 brown bakelite $75
328 painted ivory $65

AIR CASTLE

**300,325, c.1947**
5 tubes, 1 band
300 brown bakelite $50
325 ivory plaskon $65

AIR CASTLE

**7000,7200, c.1939**
7 tubes, 2 bands, AC/DC
$125

AIR CASTLE

**7002, c.1939**
5 tubes, 2 bands
$65

AIR CASTLE

**7004, c.1939**
8 tubes, 3 bands
$110

AIR CASTLE

**7008, c.1939**
4 tubes, 1 bands, DC
$30

AIR CASTLE

**7100, c.1939**
6 tubes, 2 bands, DC
$35

AIR CASTLE

**7104, c.1939**
5 tubes, 1 band, DC
$40

AIR CASTLE

**7108,9, c.1939**
4 tubes, 1 band, AC/DC
7108 brown bakelite $250
7109 ivory plaskon $400

AIR CASTLE

**7108D, c.1939**
4 tubes, 1 band, DC
$30

AIR CASTLE

**7110-12, c.1939**
6 tubes, 1 band
7110 brown bakelite $90
7111 ivory plaskon $125
7112 beetle plastic $275

AIREX

**4A, c.1934**
4 tubes, 1 band, AC/DC
$75

AIREX

**4B, c.1934**
4 tubes, 1 band, AC/DC
$110

AIREX

**5BB, c.1934**
5 tubes, 1 band
$225

AIREX
**5D, c.1934**
5 tubes, 2 bands, AC/DC
$85

AIREX
**5E, c.1934**
5 tubes, 2 bands, AC/DC
$125

AIREX
**5SH, c.1934**
5 tubes, 2 bands
$250

AIREX
**6C, c.1934**
6 tubes, 1 band
$200

AIREX
**BIG CHIEF, c.1933**
6 tubes, 1 band
$225

AIREX
**CHIEFTAIN, c.1933**
6 tubes, 1 band
$250

AIREX
**DUAL WAVE, c.1934**
6 tubes, 2 bands, AC/DC
$200

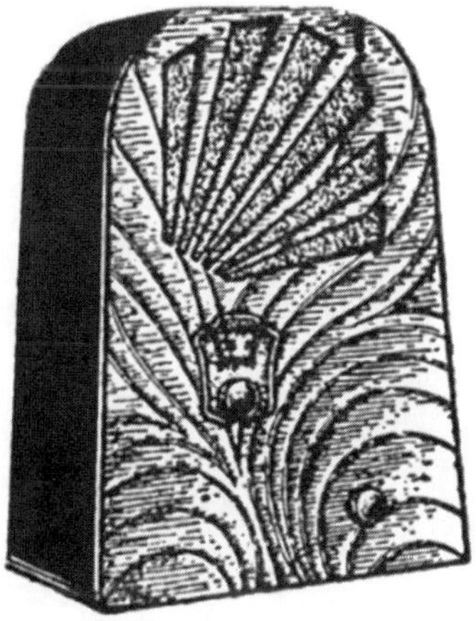

AIREX
**JUNIOR, c.1931**
4 tubes, 1 band
$350

AIREX
**MIDGET, c.1933**
5 tubes, 1 band
$200

AIREX

**MIDGET, c.1933**
6 tubes, 1 band
$375

air king

**149A, c.1937**
5 tubes, 1 band
$100

AIREX

**'MUSIQUE', c.1933**
9 tubes, 1 band
$300

air king

**'SAVIL', c.1933**
5 tubes, 1 band
$175

air king

**72, 'SKYSCRAPER', c.1933**
*Harold Van Doren Design*
5 tubes, 1 band
black $2500, crystalline $3500, brown$2000
ivory $2700, blue $6000, green $6000
lavender $6000, red $8000

air king

**222 'TUCK-IN', c.1939**
4 tubes, 1 band
brown bakelite $100
ivory plaskon $150

air king

**5000, c.1941**
6 tubes, 2 bands, radio/phono
$60

air king

**'SUPER', c.1933**
7 tubes, 1 band
$325

Airline

**5, c.1942**
5 tubes, 1 band, radio/phono
$200

Airline

**13, c.1932**
5 tubes, 1 band
$200

Airline

**25, c.1932**
6 tubes, 1 band
$250

Airline

**43, c.1932**
5 tubes, 2 bands
$250

Airline

**67, c.1933**
7 tubes, 1 band
$225

Airline

**70, c.1941**
5 tubes, 1 band
$60

Airline

**77, c.1933**
6 tubes, 1 band, DC
$150

Airline

**97, c.1933**
7 tubes, 3 bands
$200

Airline

**98, c.1933**
5 tubes, 1 band
$150

Airline

**103, c.1934**
7 tubes, 3 bands
$225

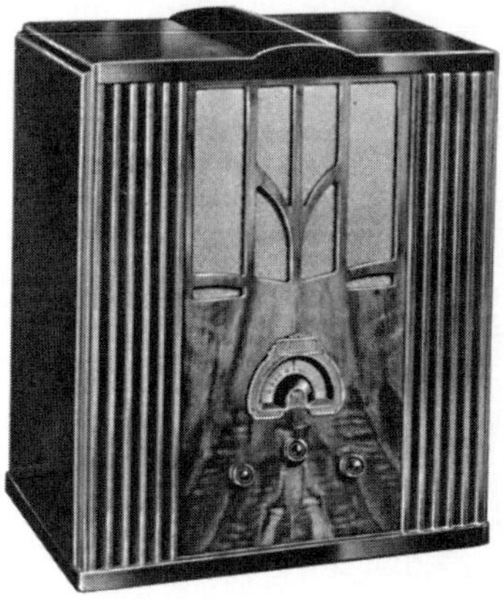

Airline

**120,141, c.1934**
5 tubes, 1 band, DC
$65

Airline

**148, c.1934**
5 tubes, 1 band
$125

Airline

**115, c.1934**
5 tubes, 1 band
$225

Airline

**117, c.1934**
5 tubes, 1 band
$225

Airline

**288,290, c.1938**
6 tubes, 1 band
288 brown bakelite $175
290 ivory plaskon $225

Airline

**422, c.1938**
7 tubes, 2 bands, DC
$225

Airline

**501,502, c.1939**
5 tubes, 1 band
501 brown bakelite $175
502 painted ivory $190

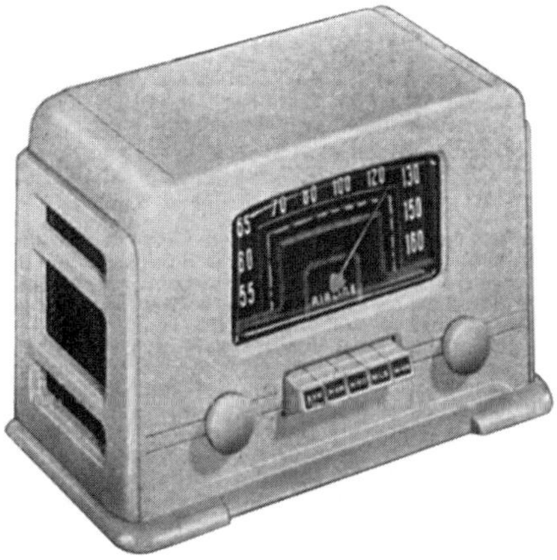

Airline

**530,531, c.1942**
5 tubes, 1 band
530 brown finish $110
531 ivory finish $125

Airline

**534,535, c.1939**
5 tubes, 1 band, DC
$35

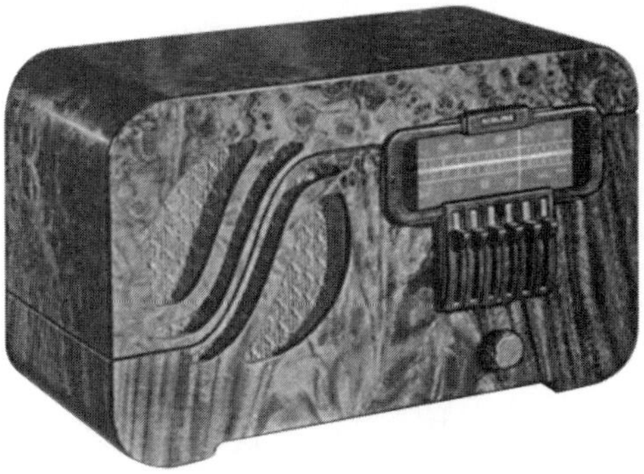

Airline

**553, c.1939**
5 tubes, 1 band
$125

Airline

**614,615, c.1941**
6 tubes, 2 bands
614 radio/phono $65
615 radio/phono/recorder $100

Airline

**722, c.1940**
7 tubes, 2 bands
$60

Airline

**754, c.1941**
7 tubes, 2 bands, DC
$30

Airline

**1522, 1931**
5 tubes, 1 band, DC
$40

**Airline**

**1523, c.1930**
5 tubes, 1 band, DC
$65 with speaker

**Airline**

**2823, c.1930**
7 tubes, 1 band, DC
$75 with speaker

**Airline**

**760 'POLICE ADAPTER', c.1932**
3 tubes, 1 band
$50

**Airline**

**5958, c.1930**
phono remote
$225

**Airline**

**1700, c.1931**
4 tubes, 1 band
$250

**ATWATER KENT**

**415Q, 435, c.1935**
415Q 5 tubes, 1 band, DC $150
435 5 tubes, 2 bands $225

**ALLEN-HOUGH**

**c.1935**
coin-operated phonograph
$225

**AMERICAN-BOSCH**

**Motorboat Radio, c.1932**
5 tubes, 1 band, DC
$125

**Andrea**

**14E6, c.1939**
6 tubes, 2 bands
$275

**Andrea**

**826, c.1940**
8 tubes, 3 bands
$325

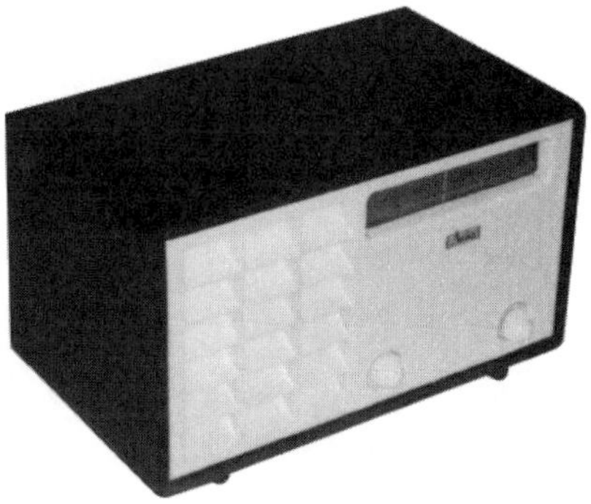

ARIA

**571, c.1946**
5 tubes, 1 band
$50

**ARKAY**

**421, c.1935**
4 tubes, 1 band
$250

**ARKAY**

**426, c.1935**
4 tubes, 1 band
$125

**ARKAY**

**521, c.1935**
5 tubes, 1 band
$175

**ARKAY**

**631, c.1935**
6 tubes, 3 bands
$150

**ARKAY**

523, C.1935
5 tubes, 2 bands
$225

**ARKAY**

534, C.1935
5 tubes, 2 bands
$200

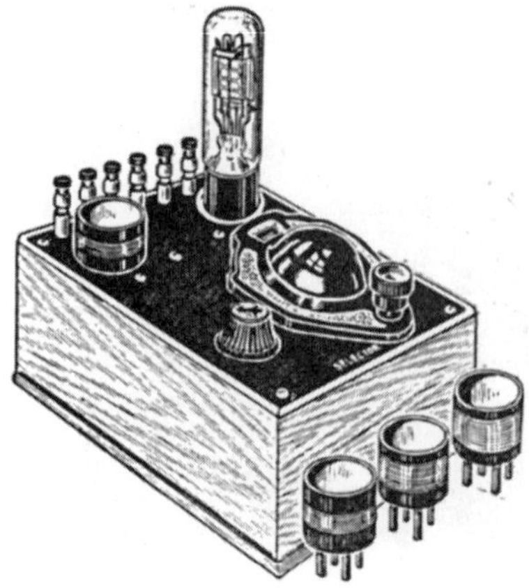

**AROUND-THE-WORLD**

**SW RECEIVER, C.1932**
1 tube, shortwave coils
$150

ARROW

**C2V5, c.1938**
5 tubes, 1 band, DC
$25

ARROW

**C18, c.1938**
8 tubes, 3 bands
$150

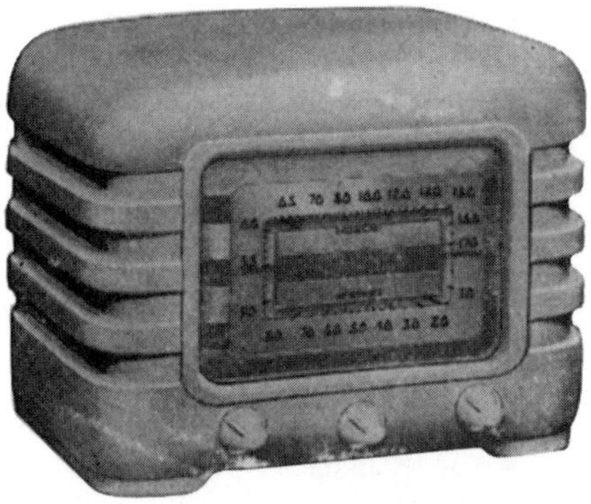

ARROW

**C57,58, c.1938**
5 tubes, 2 bands, AC/DC
C57 black bakelite $225
C58 ivory plaskon $275

ARROW

**C824, c.1938**
8 tubes, 3 bands
$250

ARROW

**T22, c.1938**
6 tubes, 3 bands
$75

ARROW

**T44, c.1938**
4 tubes, 1 band, AC/DC
$75

ARROW
**T52, c.1938**
5 tubes, 2 bands
$110

ARROW
**T62, c.1938**
6 tubes, 2 bands
$80

ARROW
**T63, c.1938**
6 tubes, 3 bands
$125

ARROW
**T83, 1938**
8 tubes, 3 bands
$175

**ARVIN**
**88, c.1939**
6 tubes, 1 band, radio/phono
$75

**ARVIN**
**89, c.1941**
6 tubes, 1 band, radio/phono
$50

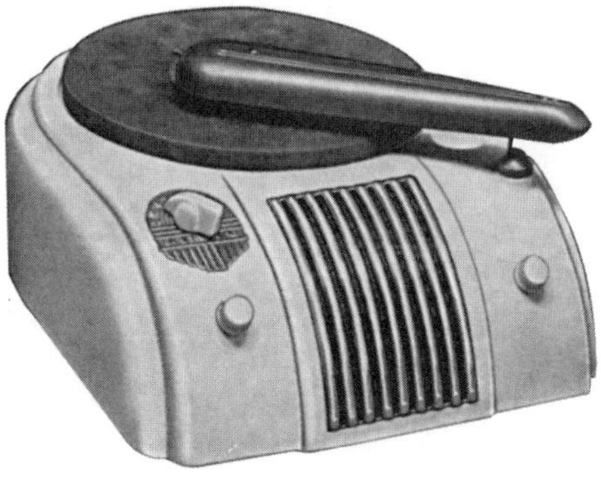

**ARVIN**
**302,302A, c.1940**
3 tubes, 1 band, radio/phono
302 brown finish $225
302A ivory finish $250

**ARVIN**
**402,402A, c.1941**
4 tubes, 1 band
402 brown finish $125
402A ivory finish $150

**ARVIN**
**40A, c.1939**
2 tubes, 1 band
ivory finish $250

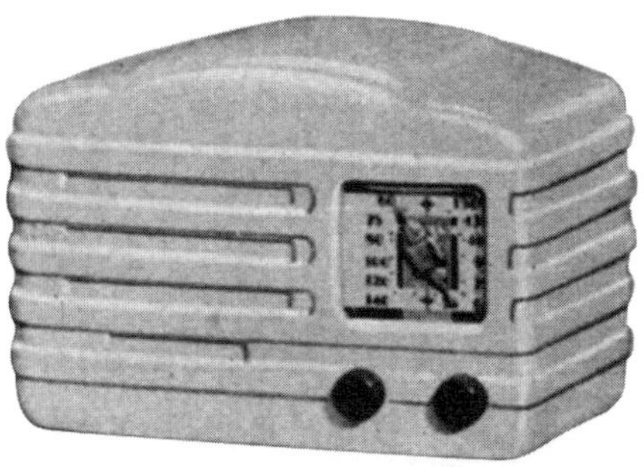

**ARVIN**
**524,524A, c.1941**
5 tubes, 1 band
524 brown finish $110
524A ivory finish $125

**ARVIN**
**524, c.1946**
5 tubes, 1 band, AC/DC
$40

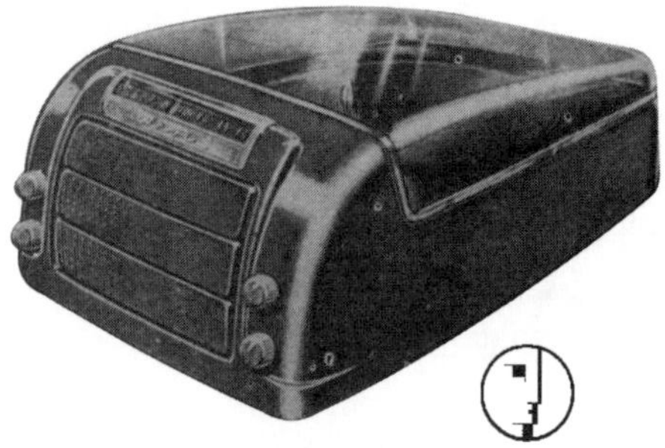

**ARVIN**
**526, c.1946**
Karstadt Design
5 tubes, 1 band, radio/phono
$110

**ARVIN**
**541B, c.1946**
4 tubes, 1 band, DC
$20

**ARVIN**
**555, c.1946**
5 tubes, 1 band, AC/DC
$30

**ARVIN**
**558, c.1946**
5 tubes, 1 band, radio/phono
$30

**ARVIN**
**559, c.1946**
5 tubes, 1 band, radio/phono
$35

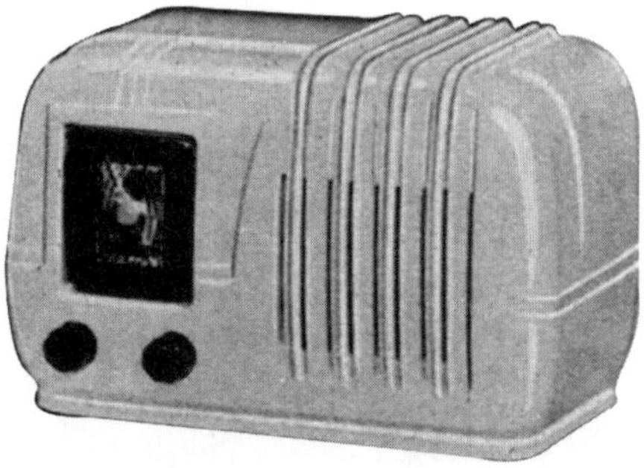

**ARVIN**
**616,616A, c.1942**
6 tubes, 1 band, AC/DC
616 brown bakelite $90
616A ivory finish $110

**ARVIN**
**641B, c.1946**
5 tubes, 1 band, DC
$20

**ARVIN**
**655SWT, c.1953**
5 tubes, 2 bands
$30

**ARVIN**
**666, c.1946**
6 tubes, 1 bands, AC/DC
$30

**ARVIN**
**669, c.1946**
6 tubes, 1 band, radio/phono
$30

**ARVIN**
**722,722A, c.1941**
6 tubes, 1 band
722 brown bakelite $35
722A ivory finish $50

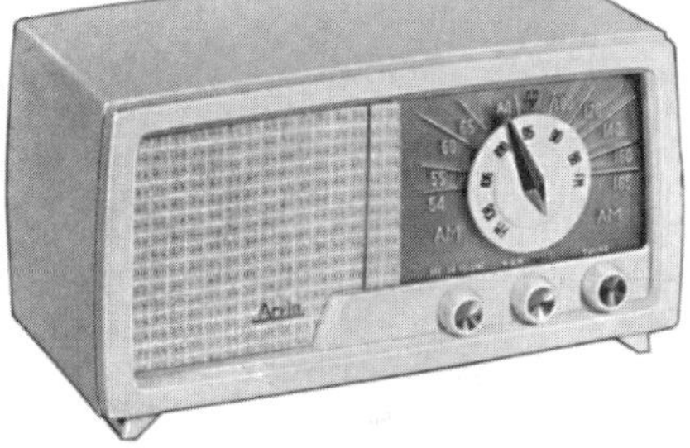

**ARVIN**
**780TFM, c.1957**
8 tubes, AM/FM
ivory $35
green $50

**ARVIN**
**950T/E/5/1-57**
5 tubes, 1 band
black $75

**ARVIN**
**242-5T, c.1950**
Karstadt Design, 4 tubes, 1 band
red $200, yellow $225, green $175, ivory $150

**ARVIN**
**250P, c.1949**
5 tubes, 1 band, AC/DC
$40

## ARVIN
**253-5T, c.1950**
Karstadt Design, 4 tubes, 1 band
black $150, brown $90, ivory $135

## ARVIN
**2564, c.1957**
5 tubes, 1 band
ivory $40, black $45

## ARVIN
**2572, c.1957**
5 tubes, 1 band
ivory $70, green $80

## ARVIN
**2573, c.1957**
5 tubes, 1 band
red $125, avocado $110

## ARVIN
**3561, c.1957**
6 tubes, 1 band
ivory $50, tan $45

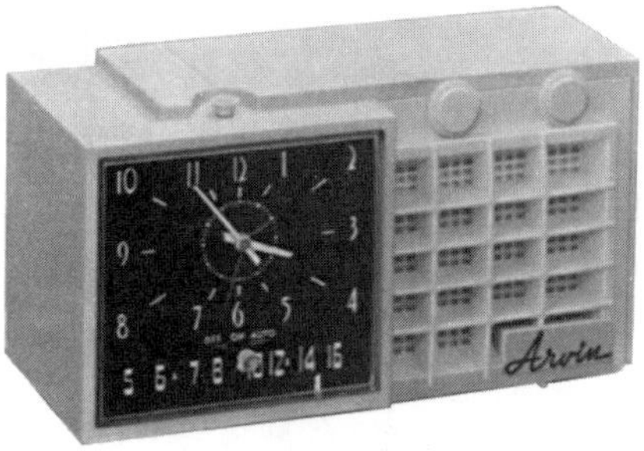


## ARVIN
**5561, c.1957**
5 tubes, 1 band
ivory $60, pink $80, maroon $70

## ARVIN
**5571, c.1957**
5 tubes, 1 band
ivory $75, green $110

## ARVIN
**5572, c.1957**
5 tubes, 1 band
red $125, black $85

## ARVIN
**5578, c.1957**
5 tubes, 1 band
ivory $60, red $95

**ARVIN**
**8571, c.1957**
5 tubes, 1 band, AC/DC
$75

**ARVIN**
**8573, c.1957**
5 tubes, 1 band, AC/DC
red $40, avocado $30

**ARVIN**
**9562, c.1957**
transistor, 1 band
$90

**ARVIN**
**9574, c.1957**
transistor, AC/DC
tan $85, ivory $90

**AUDIOLA**
**'MIDGET', c.1932**
pressboard, 4 tubes, 1 band
$275

**AUDIOLA**
**508, c.1934**
5 tubes, 1 band
$375

**AUDIOLA**
**517, c.1933**
5 tubes, 2 bands
$375

**AUDIOLA**
**810, c.1932**
8 tubes, 1 band
$425

**Autocrat**
**101, c.1939**
5 tubes, 1 band
brown bakelite $75
ivory plaskon $125
green plaskon $350

INGRAHAM

## Automatic
**402, c.1940**
5 tubes, 1 band
$150

INGRAHAM

## Automatic
**430, 'COMPANION', c.1940**
5 tubes, 1 band
$225

INGRAHAM

## Automatic
**434A, c.1940**
5 tubes, 1 band, radio/phono
$75

INGRAHAM

## Automatic
**458, c.1940**
5 tubes, 1 band
$225

## Automatic
**933 'Tom Thumb', c.1939**
metal, 5 tubes, 1 band
red, green, blue, grey, pink
$750-1,500

## Automatic
**c.1946**
5 tubes, 1 band, AC/DC
$125

## Automatic
**c.1946**
5 tubes, 1 band, radio/phono
$110

## Automatic
**c.1946**
6 tubes, 1 band
$135

## Automatic
**P25 'Tom Thumb', c.1932**
4 tubes, 1 band
$125

**Automatic**

**P45 'Tom Thumb', c.1932**
6 tubes, 1 band, clock/radio
$450

**Automatic**

**'Recordex', c.1941**
6 tubes, 2 bands, radio/phono/recorder
$150

**AZTEC**

**50, c.1932**
5 tubes, 1 band
$300

**AZTEC**

**250, c.1932**
9 tubes, 1 band
$425

**Balkeit**

**42E 'Junior', c.1933**
4 tubes, 1 band
$175

**BELMONT**

**118, c.1938**
4 tubes, 1 band
$275

**BELMONT**

**460, c.1940**
4 tubes, 1 band, DC
$25

**BELMONT**

**533, c.1940**
5 tubes, 1 band, radio/phono
$90

**BELMONT**

**586, c.1937**
5 tubes, 2 bands
$125

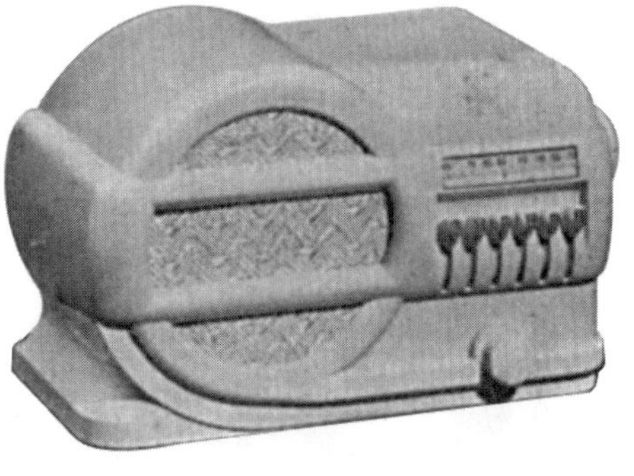

## BELMONT
**635, c.1939**
6 tubes, 1 band
$350

## BEVERLY
**'Phantom Tunette', c.1939**
remote tuner
$150

## BIRCH
**110, c.1941**
phono remote
$25

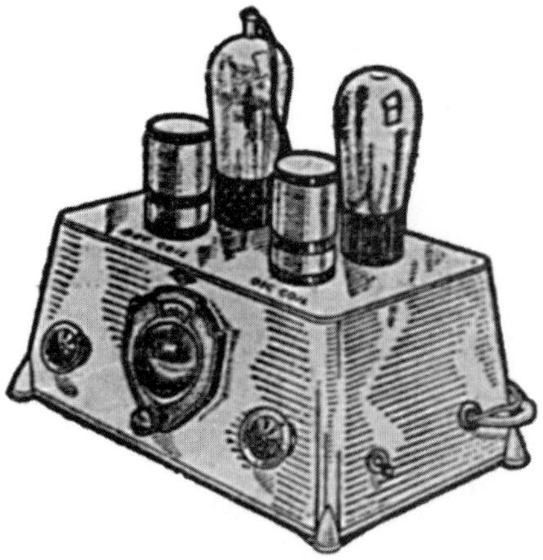

## Browning-Drake
**53,63, c.1930**
53 - 5 tubes, 1 band, $65
63 - 6 tubes, 1 band $75

## BUD
**'Super SW Converter', c.1931**
SW converter, 2 tubes
$110

## Brown & Manhart
**'Ranger', c.1931**
6 tubes, 1 band
$250

## Brunswick
**11, c.1932**
6 tubes, 1 band
$200

## Camden
**5H, c.1936**
5 tubes, 1 band
$65

**Camden**
**42 'Junior', c.1931**
5 tubes, 1 band
$175

**Camden**
**83, c.1931**
7 tubes, 1 band
$225

**Camden**
**537, c.1937**
5 tubes, 2 bands
$100

**Camden**
**606, c.1937**
7 tubes, 3 bands
$190

**Camden**
**637, c.1937**
6 tubes, 3 bands
$175

**Camden**
**837, c.1937**
8 tubes, 4 bands
$400

**Camden**
**'Minute Man', c.1932**
5 tubes, 1 band, clock/radio
$250

**Camden**
**'Multi-Mu Pentode', c.1932**
7 tubes, 1 band
$250

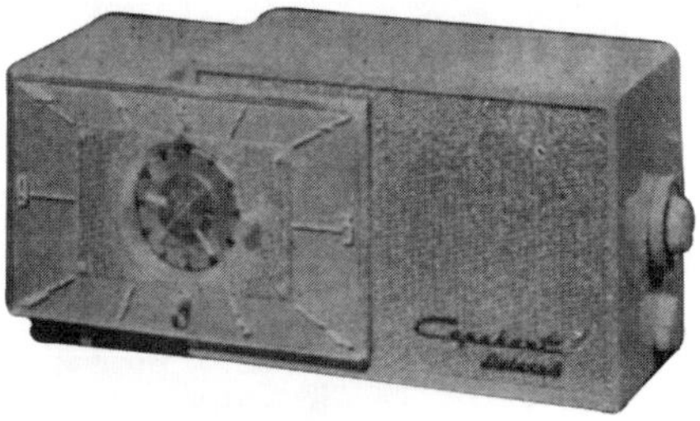

**Capehart**
**TC62 'Deluxe 6', c.1953**
6 tubes, 1 band, clock/radio
$40

## CARDINAL
**71, c.1932**
9 tubes, 1 band, radio/phono
$350

## CASE
**700, c.1936**
7 tubes, 3 bands
$250

## CASE
**710, c.1936**
7 tubes, 3 bands
$450

## CGE
CANADIAN GE
**53X, c.1937**
5 tubes, 2 bands
$125

## CGE
CANADIAN GE
**404, c.1941**
5 tubes, 1 band
$65

## CGE
CANADIAN GE
**A44, c.1936**
4 tubes, 1 band
$110

## CGE
CANADIAN GE
**A50E, c.1936**
5 tubes, 2 bands
$165

## CGE
CANADIAN GE
**A6B, c.1936**
6 tubes, 2 bands, DC
$75

## CGE
CANADIAN GE
**C400, c.1940**
4 tubes, 1 band
brown bakelite $110
ivory plaskon $150
pale blue plaskon $250

front view back view

CGE CANADIAN GE

**E50, c.1937**
5 tubes, 2 bands
$350

CGE CANADIAN GE

**F52, c.1938**
5 tubes, 2 bands
$225

CGE CANADIAN GE

**F62, c.1938**
6 tubes, 3 bands
$150

CGE CANADIAN GE

**H406U, c.1939**
4 tubes, 1 band
brown bakelite $225
ivory plaskon $350

CGE CANADIAN GE

**H61, c.1939**
6 tubes, 2 bands
$50

CGE CANADIAN GE

**JK52, c.1940**
5 tubes, 1 band
brown bakelite $125
ivory plaskon $200

CGE CANADIAN GE

**JK53, c.1940**
5 tubes, 3 bands
$225

CGE CANADIAN GE

**KL5, c.1940**
5 tubes, 4 bands, radio/phono
$75

CGE CANADIAN GE

**KL50, c.1940**
5 tubes, 1 band
$125

CGE CANADIAN GE

**KL51, c.1940**
5 tubes, 1 band
brown bakelite $75
ivory plaskon $125

CGE CANADIAN GE

**KL52, c.1940**
5 tubes, 1 band
brown bakelite $110
ivory plaskon $175

CGE CANADIAN GE

**KL53, c.1940**
5 tubes, 2 bands
$225

CGE CANADIAN GE

**KL060, c.1940**
6 tubes, 4 bands
$75

CGE CANADIAN GE

**KL70, c.1940**
7 tubes, 6 bands
$150

CGE CANADIAN GE

**KL500, c.1941**
5 tubes, 4 bands
$50

CGE CANADIAN GE

**KL501, c.1940**
5 tubes, 4 bands
$50

CGE CANADIAN GE

**R93AG, c.1940**
phono remote
$75

**CGE** CANADIAN GE

**R94G, c.1940**
phono remote
$30

**CGE** CANADIAN GE

**RW440X, c.1940**
5 tubes, 1 band
brown bakelite $110
ivory bakelite $175

**CHALLENGER**

**'Multi-Mu', c.1932**
7 tubes, 1 band
$75

*Clarion*

**402,402V,502,502V, c.1940**
4/5 tubes, 1 band
402,502 brown bakelite $75
402V,502V ivory plaskon $125

*Clarion*

**404, c.1940**
4 tubes, 1 band
$65

*Clarion*

**405, c.1940**
4 tubes, 1 band
$65

*Clarion*

**423, c.1940**
4 tubes, 1 band, DC
$40

*Clarion*

**424, c.1940**
4 tubes, 1 band, DC
$30

*Clarion*

**432, c.1940**
4 tubes, 1 band, DC
$20

Clarion
**504, c.1940**
5 tubes, 1 band
$50

Clarion
**505, c.1940**
5 tubes, 1 band
$90

Clarion
**523, c.1940**
5 tubes, 1 band, radio/phono
$40

Clarion
**602, c.1940**
6 tubes, 1 band
$75

Clarion
**603, c.1940**
6 tubes, 1 band
$90

Clarion
**612, c.1940**
6 tubes, 1 band
$50

Climax
**c.1940**
5 tubes, 1 band
$250

Climax
**D 'Diamond', c.1938**
4 tubes, 1 band
$375

Climax
**'Mini-Console', c.1937**
6 tubes, 2 band
$950

Climax
**UE64, c.1938**
7 tubes, 2 bands
$750

Climax
**UE65 'Emerald', c.1938**
7 tubes, 2 bands
$900

Climax
**UE94, c.1938**
7 tubes, 2 bands
$175

Climax
**W52L, c.1936**
6 tubes, 2 bands, DC
$325

Colonial
**'Remote Control', c.1931**
$125

Columbia
**C12, c.1932**
SW Converter
$100

Columbia
**C31, c.1932**
5 tubes, 1 band
$125

Columbia
**C33, cv.1932**
7 tubes, 1 band
$375

Columbia
**C53, c.1932**
6 tubes, 1 band
$350

Columbia
**T1,T2, c.1941**
phono remote
$25

Continental
**159-5L, c.1940**
6 tubes, 1 band, radio/phono
$85

Continental
**A682, c.1936**
6 tubes, 1 band, DC
$60

Continental
**A3282, c.1936**
6 tubes, 3 bands, DC
$70

Continental
**A6322, c.1936**
6 tubes, 3 bands
$225

Continental
**B105, c.1935**
6 tubes, 3 bands
$110

Continental
**B110, c.1935**
6 tubes, 3 bands
$250

Continental
**B125, c.1935**
6 tubes, 3 bands
$225

Continental
**B170, c.1935**
7 tubes, 3 bands
$125

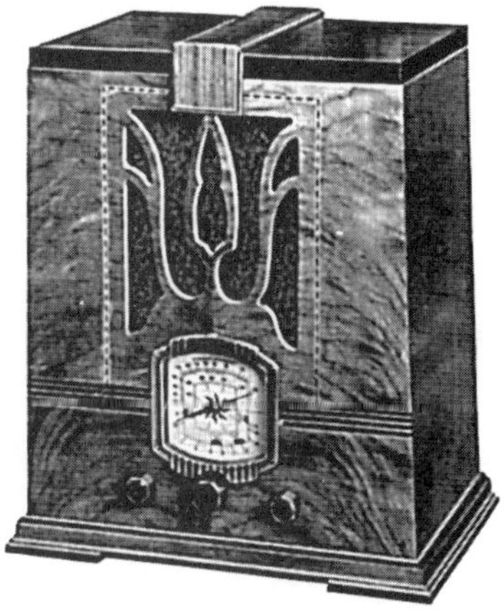

Continental
**B600, c.1935**
6 tubes, 3 bands
$150

Continental
**B632, c.1935**
5 tubes, 3 bands, DC
$40

Continental
**N523, c.1936**
5 tubes, 3 bands
$125

Continental
**N601, c.1936**
6 tubes, 3 bands
$150

CORDONIC
**Geneva, c.1931**
6 tubes, 1 band
$250

CORONA
**211, c.1937**
4 tubes, 2 bands, DC
$50

CORONA
**213, c.1937**
7 tubes, 2 bands, DC
$90

COUNTRY CLUB
**'Fisherman', c.1933**
5 tubes, 2 bands
$750

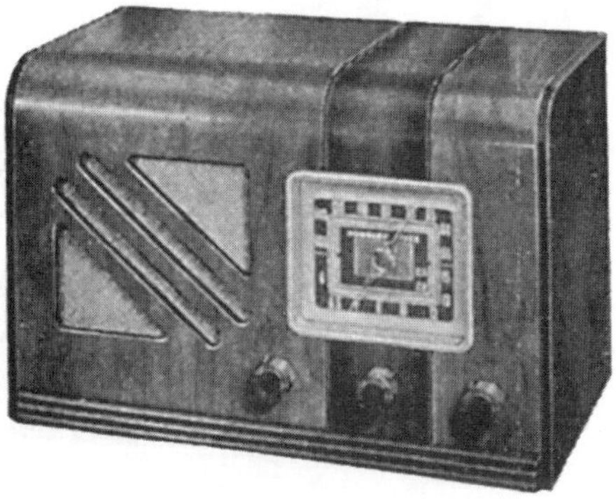

CROSLEY
**20AP, c.1941**
5 tubes, 3 bands
$50

**CROSLEY**

**13AE, c.1941**
5 tubes, 2 bands
$110

**CROSLEY**

**16AL, c.1941**
5 tubes, 1 band
$60

**CROSLEY**

**21AQ, c.1941**
6 tubes, 2 bands
$65

**CROSLEY**

**23AR, c.1941**
5 tubes, 3 bands
$60

**CROSLEY**

**24AJ, c.1941**
7 tubes, 3 bands
$110

**CROSLEY**

**31S 'Montrad', c.1930**
7 tubes, 1 band
$225

CROSLEY
**25AW, c.1941**
7 tubes, 3 bands
$175

CROSLEY
**32BG, c.1941**
6 tubes, 1 band, radio/phono/recorder
$110

CROSLEY
**52TD, c.1941**
5 tubes, 2 bands
$110

CROSLEY
**52TE, c.1941**
5 tubes, 2 bands
$125

CROSLEY
**52TF, c.1941**
5 tubes, 2 bands
$60

CROSLEY
**52TG, c.1942**
5 tubes, 2 bands
$110

CROSLEY
**52TP, c.1941**
5 tubes, 2 bands, radio/phono
$60

CROSLEY
**56TH, c.1941**
5 tubes, 1 band
$100

CROSLEY
**62TA, c.1942**
6 tubes, 2 bands
$80

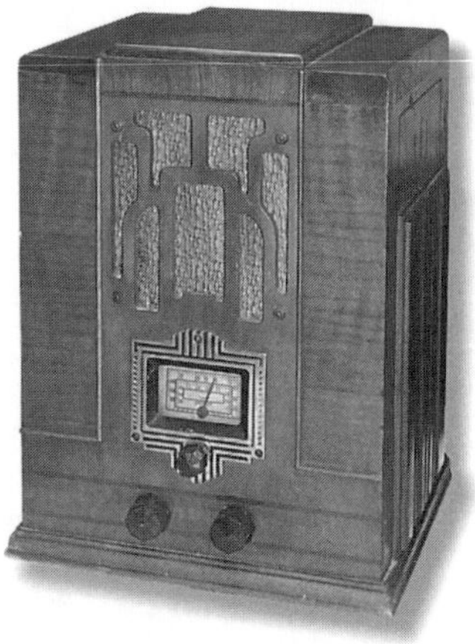

**CROSLEY**
**5V1, c.1933**
5 tubes, 1 bands
$225

**CROSLEY**
**62TC, c.1942**
6 tubes, 2 bands
$85

**CROSLEY**
**62TD, c.1942**
6 tubes, 2 bands
$60

**CROSLEY**
**159, c.1933**
5 tubes, 1 band
$175

**CROSLEY**
**415AA, 'Battery 4', c.1934**
4 tubes, 1 band
$150

**CROSLEY**
**515AC 'Fiver', c.1936**
5 tubes, 2 bands
$125

**CROSLEY**
**519A,529A, c.1940**
5 tubes, 1 bands
$175

**CROSLEY**
**545AD 'Privateer', c.1935**
5 tubes, 1 band
$110

**CROSLEY**
**555KC 'Battery 5', c.1933**
5 tubes, 2 bands, DC
$175

**CROSLEY**
**529D, c.1940**
5 tubes, 1 band
$150

**CROSLEY**
**601 'Bandbox', c.1931**
6 tubes, 1 band
$75

**CROSLEY**
**599A, c.1940**
5 tubes, 1 band
$150

**CROSLEY**
**0615C 'Cruiser', c.1936**
6 tubes, 3 bands
$150

**CROSLEY**
**625E 'Battery 6', c.1936**
6 tubes, 3 bands, DC
$125

**CROSLEY**
**635C 'Buccaneer', c.1936**
6 tubes, 3 bands
$150

**CROSLEY**
**637, c.1938**
6 tubes, 2 bands
$70

**CROSLEY**
**649A, c.1940**
5 tubes, 1 band
$200

**CROSLEY**
**645CB '32 VOLT 6', c.1936**
6 tubes, 2 bands, DC
$100

**CROSLEY**
**649D, c.1940**
5 tubes, 1 band
$175

**CROSLEY**
**715D 'Corsair', c.1936**
7 tubes, 3 bands
$150

**CROSLEY**
**719A, c.1940**
7 tubes, 2 bands
painted finish: red $125, ivory $100,
tan $110, blue $125

**CROSLEY**
**719D, c.1940**
7 tubes, 2 bands
$100

**CROSLEY**
**725F 'Viking', c.1936**
7 tubes, 5 bands
$160

**CROSLEY**
**739A, c.1940**
7 tubes, 2 bands
$110

**CROSLEY**
**855D 'Merrimac', c.1936**
8 tubes, 3 bands
$275

**CROSLEY**
**865F 'Monitor', c.1936**
8 tubes, 5 bands
$250

**CROSLEY**
**815EC, 915EK, 1055EK, c.1936**
815EC 'Battery 8', 8 tubes, 2 bands, $175
915EK 'Clipper', 9 tubes, 5 bands, $425
815EC 'Constitution', 10 tubes, 5 bands, $500

**CROSLEY**
**AFM, c.1936**
5 tubes, 2 bands
$145

**CROSLEY**
**B449A, c.1940**
4 tubes, 1 band, DC
$75

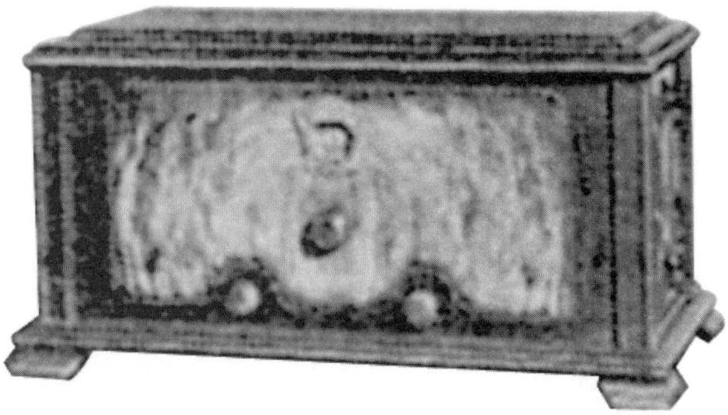

**CROSLEY**
**'Alderman', c.1932**
6 tubes, 1 band
$125

**CROSLEY**
**B459A, c.1940**
4 tubes, 1 band, DC
$125

**CROSLEY**
**B5579M,B5589M, c.1940**
5 tubes, 1/2 bands, DC
$125

**CROSLEY**
**B579A, c.1940**
5 tubes, 1 band, DC
$50

**CROSLEY**
**B589A, c.1940**
5 tubes, 2 bands, DC
$50

**CROSLEY**
**'Bookcase', c.1933**
5 tubes, 1 band
$250

**CROSLEY**
**C33-CA, c.1941**
6 tubes, 3 bands, radio/phono/recorder
$110

**CROSLEY**
**'Dual Fiver', c.1934**
5 tubes, 2 bands
$175

**CROSLEY**
**'Congressman', c.1933**
9 tubes, 1 band, DC
$175

**CROSLEY**
**'Fiver', c.1938**
5 tubes, 1 band, AC/DC
$60

**CROSLEY**
**'Fiver Compact', c.1938**
5 tubes, 2 bands
$110

**CROSLEY**
**'Fiver Deluxe', c.1938**
5 tubes, 2 bands
$75

**CROSLEY**
**'Fiver Deluxe Compact', c.1938**
5 tubes, 2 bands
$60

CROSLEY
**'Fivette', c.1938**
5 tubes, 1 band
$125

CROSLEY
**'Grand Station', c.1934**
7 tubes, 2 bands
$375

CROSLEY
**'Littlfella', c.1932**
5 tubes, 1 band
$225

CROSLEY
**'New Pup', c.1933**
4 tubes, 1 band
$75

CROSLEY
**'Nomad', c.1932**
5 tubes, 1 band
$60

CROSLEY
**'Mayor', c.1933**
9 tubes, 1 band
$325

CROSLEY
**'Super 6', c.1938**
6 tubes, 2 bands
$75

CROSLEY
**'Super 8', c.1938**
8 tubes, 3 bands
$90

**CROSLEY**
**'Vagabond', c.1933**
5 tubes, 1 band
$60

*Cruzair*
**703, c.1941**
7 tubes, 2 bands
$50

**DECCA**
**c.1939**
phono remote
$60

DEFOREST
**'Royal', c.1931**
7 tubes, 1 band
$275

DEFOREST
**'Aires', c.1937**
5 tubes, 1 band
$110

DEFOREST
**'Belle Isle', c.1937**
9 tubes, 3 bands
$325

DEFOREST
**'Dufferin', c.1938**
7 tubes, 3 band
$225

DEFOREST
**'Erie', c.1938**
6 tubes, 2 bands
$350

**DELCO**
**M Series 'Yacht', c.1940**
9 tubes, 2 bands
$500

DE LUXE
**'Mayfair'c.1934**
4 tubes, 1 band
$175

DE LUXE
**'Pal', c.1934**
5 tubes, 1 band
$250

DE LUXE
**25, c.1935**
5 tubes, 1 band, TRF
$150

DE LUXE
**36 'Allwave Super', c.1937**
5 tubes, 3 bands
$350

DETROLA
**100C, c.1935**
6 tubes, 2 bands
$450

DETROLA
**104, c.1935**
4 tubes, 1 bands
$400

DETROLA
**114, c.1936**
6 tubes, 3 bands
$350

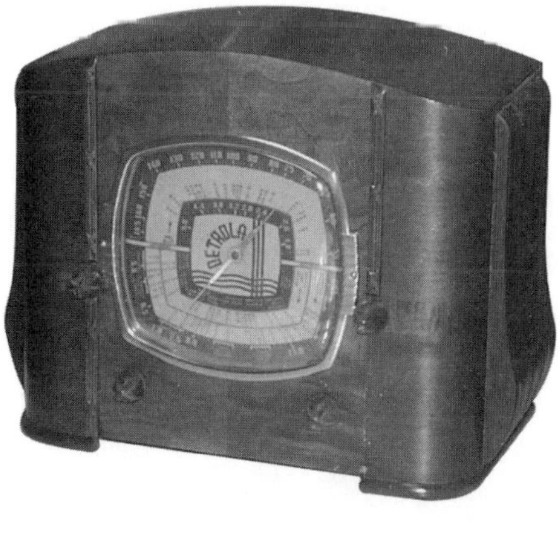

DETROLA
**140, c.1937**
6 tubes, 2 bands
$325

DETROLA
**162, c.1935**
6 tubes, 2 bands
$150

**DETROLA**
**172A 'GLEN', c.1938**
5 tubes, 1 band
$200

**DETROLA**
**221, c.1939**
6 tubes, 2 bands
$250

**DETROLA**
**222, c.1939**
6 tubes, 2 bands
$325

**DETROLA**
**1137, c.1937**
6 tubes, 3 bands
$225

**DETROLA**
**2811 'Split Grille', c.1939**
5 tubes, 1 band
$375

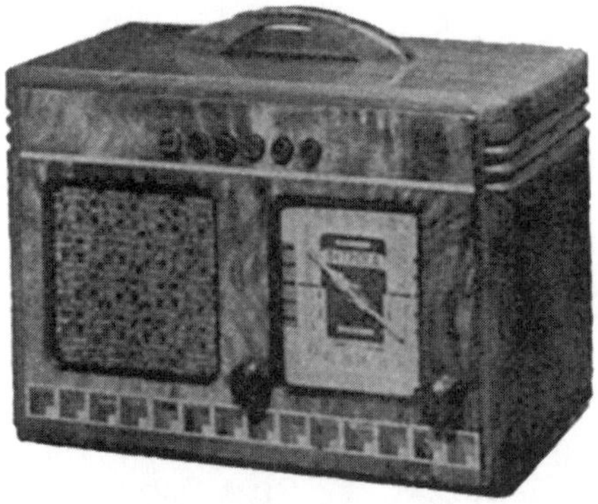

**DETROLA**
**3051, c.1940**
7 tubes, 2 bands
$250

**DETROLA**
**4D1, c.1934**
4 tubes, 1 band
$450

**DETROLA**
**CM254A 'Bullet', c.1940**
6 tubes, 2 bands
$350

**DETROLA**
**'Midget', c.1932**
5 tubes, 1 band
$250

**DEWALD**
**60-3, c.1933**
6 tubes, 1 band
$100

**DEWALD**
**60A, 503A, c.1935**
60A 6 tubes, 2 bands $165
503A 5 tubes, 2 bands $150

**DEWALD**
**406 'Bantam', c.1939**
4 tubes, 1 band
brown bakelite $275
ivory plaskon $350

**DEWALD**
**411, c.1941**
phono remote
$50

INGRAHAM

**DEWALD**
**414, c.1933**
4 tubes, 1 band
$250

**DEWALD**
**440, c.1934**
repwood, 4 tubes, 1 band
$450

**DEWALD**
**500, c.1940**
5 tubes, 1 band
$40

**DEWALD**
**531,532, c.1939**
5 tubes, 1 band, radio/phono
$75

**DEWALD**
**533, c.1939**
5 tubes, 1 band
$125

**DEWALD**

**534, c.1939**
5 tubes, 1 band
$80

**DEWALD**

**537,537SW, c.1940**
5 tubes, 1/2 bands
$90

**DEWALD**

**538L,538SW, c.1940**
5 tubes, 1/2 band
brown bakelite $125
ivory plaskon $200

**DEWALD**

**542/662, c.1940**
5/6 tubes, 2 bands
$50

**DEWALD**

**548/549/555, c.1941**
5 tubes, 2/3 bands
brown bakelite $300
ivory plaskon $500

INGRAHAM

**DEWALD**

**551, c.1933**
5 tubes, 1 band
$275

**DEWALD**

**51, c.1933**
5 tubes, 1 band
$100

**DEWALD**

**556,558, c.1941**
5 tubes, 2/3 bands
$35

**DEWALD**

**570, c.1934**
5 tubes, 1 band
$375

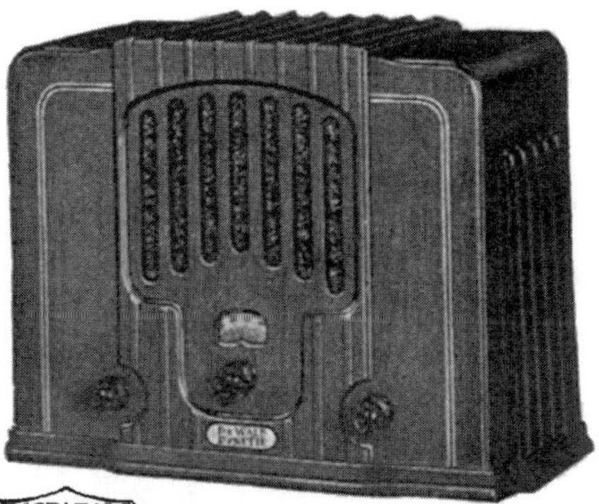

INGRAHAM

**DEWALD**

**580, c.1934**
5 tubes, 1 band
$450

**DEWALD**

**610SA, c.1936**
6 tubes, 3 bands
$250

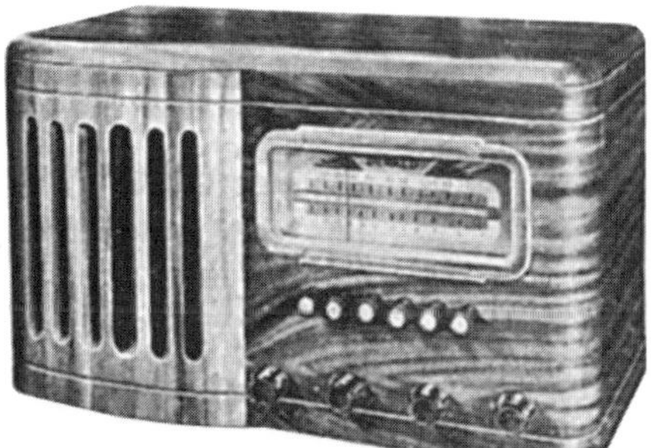

**DEWALD**

**643, c.1939**
6 tubes, 2 bands
$110

**DEWALD**

**645, c.1939**
6 tubes, 2 bands
$125

**DEWALD**

**648,650, c.1939**
7 tubes, 2/3 bands
$90

**DEWALD**

**649,651, c.1939**
6 tubes, 2/3 bands
$125

**DEWALD**

**653,654, c.1940**
6 tubes, 2 bands, radio/phono
$60

**DEWALD**

**660, c.1940**
6 tubes, 3 bands
$110

**DEWALD**

**663,664, c.1940**
6 tubes, 1 band, radio/phono
$50

**DEWALD**
**705, c.1940**
7 tubes, 3 bands
$70

**DEWALD**
**802A, c.1935**
8 tubes, 4 bands
$125

**DEWALD**
**810,812, c.1940**
8 tubes, 3 bands
$175

**DEWALD**
**811R, c.1934**
8 tubes, 3 bands
$250

**DEWALD**
**908, c.1941**
9 tubes, 3 bands, radio/phono/recorder
$250

**DEWALD**
**1004, c.1940**
10 tubes, 4 bands
$225

**DEWALD**
**5380,5380SW, c.1941**
5 tubes, 1/2 bands
brown bakelite $100
beetle plastic $350

**DEWALD**
**A503R 'Streamliner', c.1946**
5 tubes, 1 band
$25

**DEWALD**
**A504 'Excel', c.1946**
5 tubes, 1 band
$35

**DEWALD**
**A605 'Connoisseur', c.1946**
6 tubes, 1 band, radio/phono
$30

**DEWALD**
**B401 'Post-War Bantam'c.1948**
4 tubes, 1 band
$225

**DEWALD**
**B403 'Harp', c.1948**
5 tubes, 1 band, clock/radio
Catalin: ivory $600, maroon $750
brown $650

**DEWALD**
**B500 'Leader', c.1948**
5 tubes, 1 band
$25

**DEWALD**
**B504, c.1948**
4 tubes, 1 band, AC/DC
$125

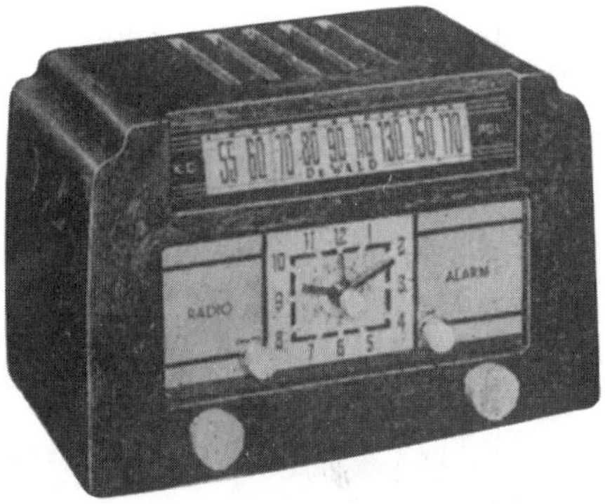

**DEWALD**
**B504, c.1948**
5 tubes, 1 band, clock/radio
Catalin: ivory $450, maroon $500
brown $550

**DEWALD**
**B612, c.1948**
Wireless FM, 8 tubes
$40

**DIETZEN**
**38-2, c.1939**
6 tubes, 1 band
$50

**DIETZEN**
**252, c.1939**
6 tubes, 1 band
$125

**DIETZEN**

**257, c.1939**
6 tubes, 1 band
$75

**DIETZEN**

**900, c.1939**
5 tubes, 1 band
$50

**DIETZEN**

**910, c.1939**
5 tubes, 1 band
$60

**DIETZEN**

**912, c.1939**
6 tubes, 1 band
$125

**DIETZEN**

**914, c.1939**
6 tubes, 2 bands
$125

**DIETZEN**

**920, c.1939**
10 tubes, 3 bands
$210

**DIETZEN**

**930, c.1939**
4 tubes, 1 band
$350

**DIETZEN**

**950, c.1939**
5 tubes, 1 band
$450

**DIETZEN**

**960, c.1939**
6 tubes, 1 band
$75

Echophone

**40 'Echoette', c.1932**
4 tubes, 1 band
$250

Echophone

**60, c.1932**
6 tubes, 1 band
$475

Echophone

**80, c.1932**
8 tubes, 1 band
$500

Electromatic

**607A 'Tank', c.1946**
Artek Pascoe Design
6 tubes, 1 band
$250

Electrohome

**5C12, c.1956**
5 tubes, 1 band,clock/radio
yellow $80, pink $70

Electrohome

**5C12, c.1956**
5 tubes, 1 band
yellow $80, pink $70

Emerson

**51, c.1932**
4 tubes, 2 bands
$175

Emerson

**52, c.1932**
5 tubes, 2 bands
$225

Emerson

**53, c.1932**
5 tubes, 2 bands
$250

**Emerson**

**108, c.1936**
6 tubes, 3 bands, clock/radio
$1500

**Emerson**

**236 'Snow White', c.1939**
repwood, 4 tubes, 1 band
$2500

**Emerson**

**250, c.1932**
5 tubes, 1 band
$175

**Emerson**

**376, c.1941**
5 tubes, 1 band
$275

**Emerson**

**382, c.1940**
5 tubes, 1 band, radio/phono
$35

**Emerson**

**459, c.1933**
4 tubes, 1 band, DC
$275

**Emerson**

**505, c.1946**
5 tubes, 1 band, AC/DC
$50

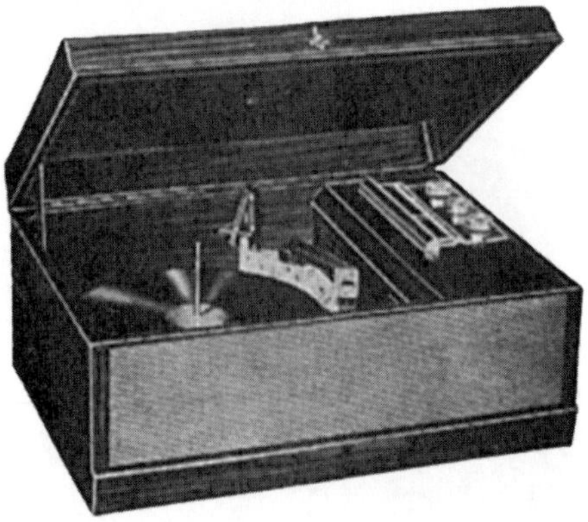

**Emerson**

**506, c.1947**
5 tubes, 1 band, radio/phono
$275

**Emerson**

**508, c.1947**
4 tubes, 1 band, DC
$110

**Emerson**

**525, c.1947**
5 tubes, 1 band, radio/phono
$75

**Emerson**

**546, c.1947**
5 tubes, 1 band, radio/phono
$45

**Emerson**

**559 'RadioChest', c.1933**
5 tubes, 1 band
$350

**Emerson**

**561A, c.1947**
Raymond Loewy Attrib.
5 tubes, 1 band, radio/phono
$150

**Emerson**

**584, c.1952**
4 tubes, 1 band
$75

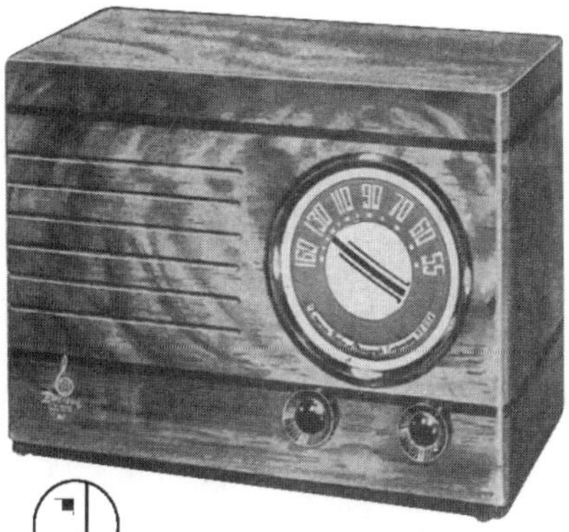

**Emerson**

**594, c.1949**
Norman Bel Geddes Design
5 tubes, 1 band
$175

**Emerson**

**596, c.1949**
5 tubes, 1 band, radio/phono
$35

**Emerson**

**640, c.1952**
4 tubes, 1 band
$110

**Emerson**

**656, c.1952**
6 tubes, 1 band, AC/DC
$75

**Emerson**

**695, c.1952**
5 tubes, 1 band, clock/radio
brown $50, ivory $60, red $125

**Emerson**

**702, c.1952**
5 tubes, 1 band
$45

**Emerson**

**657, c.1952**
6 tubes, 1 band, AC/DC
$45

**Emerson**

**703, c.1952**
5 tubes, 1 band, radio/phono
$40

**Emerson**

**805B, c.1954**
5 tubes, 1 band
brown $40, ivory $50

**Emerson**

**805B, c.1954**
5 tubes, 1 band, AC/DC
brown $75, ivory $110, orange $150

**Emerson**

**850, c.1952**
4 tubes, 1 band
maroon $100, grey $100, green $110
black $110, pink $135, yellow $150

**EMPIRE**

**25, c.1933**
4 tubes, 1 band
$275

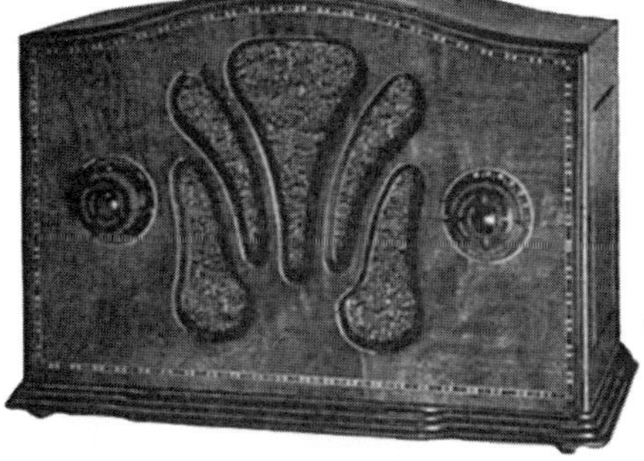

## EMPIRE
**30, c.1933**
5 tubes, 1 band
$175

## EMRO
**593 'Four', c.1933**
4 tubes, 1 band
$150

## EMRO
**627 'Tea Chest', c.1933**
4 tubes, 1 band
$300

## EMRO
**628 'Senior 6', c.1933**
6 tubes, 1 band
$175

## EMRO
**629 'Junior', c.1933**
4 tubes, 1 band
$250

## ERLA
**22P, c.1932**
4 tubes, 1 band, clock/radio
$750

## ERLA
**Shortwave Converter, c.1932**
3 tubes
$150

## ERLA
**'Variable-Mu', c.1932**
4 tubes, 1 band
$450

## ESPEY
**AWC, c.1941**
phono only
$20

FADA Radio

**6A51I,6A51W, c.1939**
5 tubes, 2 bands
brown $175, ivory $225

FADA Radio

**20T, c.1938**
6 tubes, 1 band
$125

FADA Radio

**34, c.1941**
5 tubes, 1 band
$125

FADA Radio

**55, c.1933**
5 tubes, 1 band
$275

FADA Radio

**158J,158JK,190T, c.1935**
158J - 8 tubes, 2 bands $325
158J - 8 tubes, 2 bands, Hi-Fi $500
190T - 9 tubes, 4 bands $450

FADA Radio

**F55W, F55I, c.1939**
5 tubes, 1 band
brown $125, ivory $175

FADA Radio

**'Fadalette', c.1933**
5 tubes, 1 band
pressboard: ivory $150, green $175
red $225, brown $125

FADA Radio

**PL50, c.1940**
5 tubes, 1 band, radio/phono
$65

**FAIRBANKS MORSE**

**7A, c.1941**
6 tubes, 3 bands
$225

**FAIRBANKS MORSE**

**57TO, c.1937**
5 tubes, 2 bands
$150

**FAIRBANKS MORSE**

**72T3, c.1937**
7 tubes, 3 bands
$275

**FAIRBANKS MORSE**

**91T4, c.1937**
9 tubes, 4 bands
$750

*Farnsworth*

**52,53, c.1942**
5 tubes, 1 bands
52 - brown $25
53 - ivory $35

*Farnsworth*

**AK17, c.1940**
5 tubes, 1 band, radio/phono
$75

*Farnsworth*

**BKR84, c.1941**
8 tubes, 2 bands, radio/phono/recorder
$90

*Farnsworth*

**BP2, c.1941**
phono remote
$20

*Farnsworth*

**BT40, c.1940**
4 tubes, 1 band, DC
$20

*Farnsworth*

**CT50,51, c.1942**
5 tubes, 1 band
CT50 brown $25
CT51 ivory $30

*Farnsworth*

**CT54, c.1942**
5 tubes, 1 band
$40

*Farnsworth*

**CT59 'Caryette', c.1941**
5 tubes, 1 band
$50

*Farnsworth*

**CT61,62, c.1942**
6 tubes, 1 band
CT61 brown $25
CT62 ivory $30

*Farnsworth*

**CT64, c.1942**
5 tubes, 2 bands, bandspread
$50

*Farnsworth*

**EP3505T, c.1946**
5 tubes, 1 band, AC/DC
$75

*Farnsworth*

**ET51,52, c.1946**
6 tubes, 1 band
ET51 black $50
ET52 ivory $50

*Farnsworth*

**ET55, c.1946**
5 tubes, 1 band
$30

*Farnsworth*

**ET56, c.1946**
5 tubes, 1 band
$35

*Farnsworth*

**ET63, c.1946**
5 tubes, 1 band
$35

*Federal*

**101, c.1940**
radio/phono/recorder
$90

*Federal*

**1025F, c.1947**
5 tubes, 2 band, AC/DC
$25

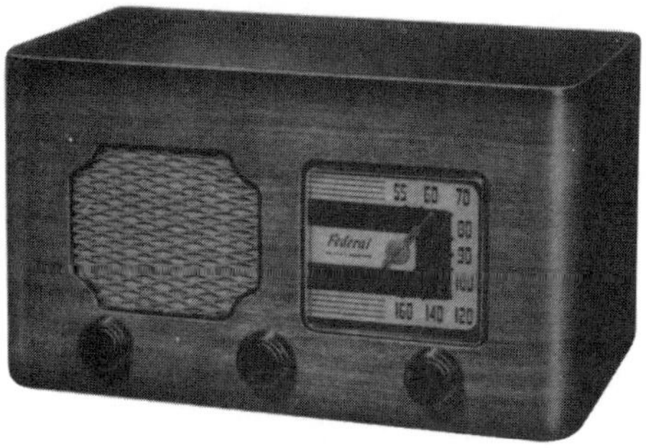


*Federal*

**1028, c.1947**
5 tubes, 1 band
$25

*Federal*

**1029, c.1947**
5 tubes, 1 band
aqua $40, ivory $30

*Federal*

**1040 Deluxe, c.1947**
5 tubes, 1 band
aqua & gold $90, ivory & gold $75
black & gold $75

*Federal*

**1040, c.1947**
5 tubes, 1 band
brown $40, black $60, ivory $75

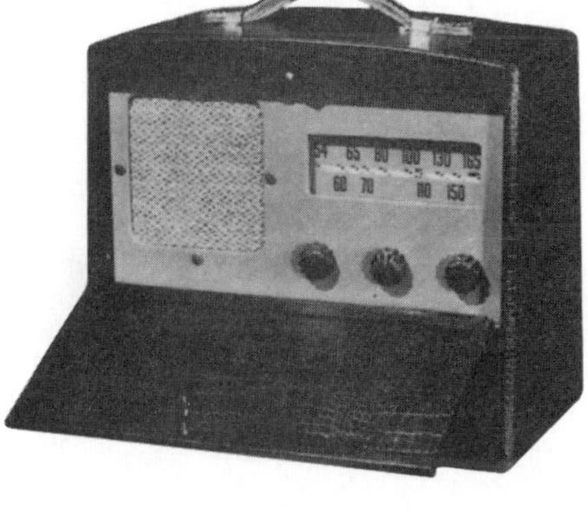

*Federal*

**6001PO, c.1947**
5 tubes, 1 band, AC/DC
$20

*Federal*

**'Allwave 6', c.1938**
6 tubes, 3 bands
$125

*Federal*

**'Allwave 8', c.1938**
8 tubes, 3 bands
$225

*Federal*

**'Five-Tube TRF', c.1938**
5 tubes, 2 bands
$150

*Federal*

**'Manual-Matic', c.1938**
5 tubes, 1 band
$375

*Federal*

**'Midget TRF', c.1938**
4 tubes, 1 band
natural $100, ivory finish $125

*Federal*

**'Superhet 6', c.1938**
6 tubes, 2 bands
$125

**Firestone**

**4A2, c.1947**
5 tubes, 1 band
brown $35, ivory $40

**Firestone**

**4A20, c.1947**
6 tubes, 2 bands
$30

**Firestone**

**4A21,22, c.1947**
6 tubes, 2 bands
$90

**Firestone**

**4A24, c.1947**
4 tubes, 1 band, DC
$20

**Firestone**

**S7397-1, c.1941**
5 tubes, 1 band, AC/DC
$125

**Firestone**
**S7398-1, c.1941**
5 tubes, 1 band
$135

**Firestone**
**S7401-8, c.1940**
John Vassos Design
5 tubes, 1 band
$225

**Firestone**
**S7402-5, c.1941**
5 tubes, 1 band
brown $30, ivory $40

**Firestone**
**S7402-7, c.1941**
5 tubes, 1 band
brown $75, ivory $90

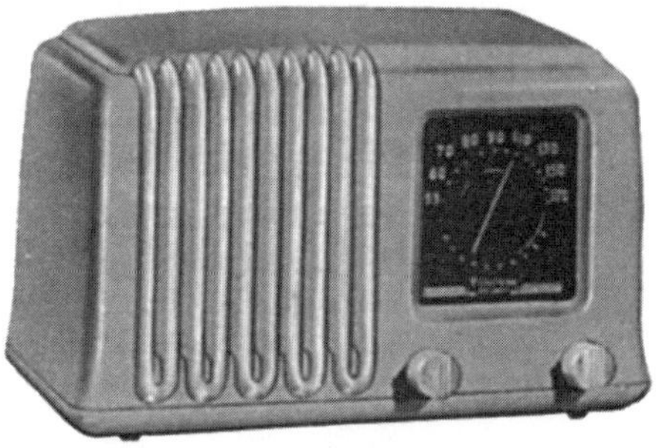

**Firestone**
**S7403-2, c.1941**
6 tubes, 1 band
brown $90, ivory $110

INGRAHAM

**Firestone**
**S7403-6, c.1941**
6 tubes, 1 band
$375

INGRAHAM

**Firestone**
**S7403-7, c.1941**
6 tubes, 2 bands
$350

INGRAHAM

**Firestone**
**S7403-9, c.1941**
8 tubes, 3 bands
$175

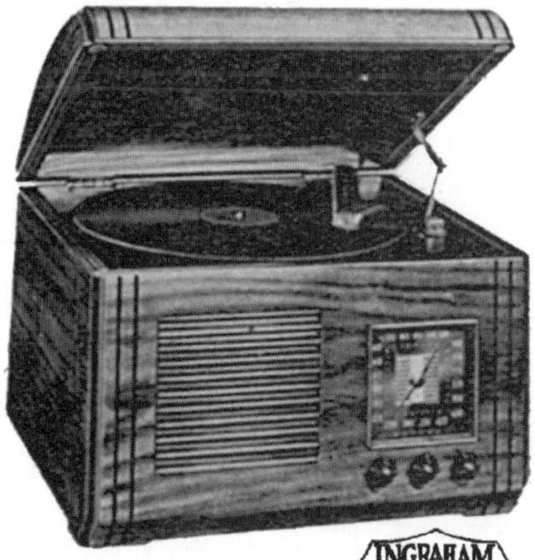

INGRAHAM

**Firestone**
**S7406-5, c.1941**
6 tubes, 2 bands, radio/phono
$125

**Firestone**

**S7428-2, c.1941**
4 tubes, 1 band, DC
$85

**FLINT**

**'Superhet', c.1931**
8 tubes, 1 band
$400

**FRANKLIN**

**'Old Standby', c.1933**
5 tubes, 1 band
$135

*Freed-Eisemann*

**55, c.1933**
5 tubes, 1 band
$75

*Freed-Eisemann*

**55P, c.1933**
5 tubes, 1 band
$75

*Freed-Eisemann*

**72, c.1933**
7 tubes, 1 band
$225

*Freed-Eisemann*

**77, c.1933**
7 tubes, 1 band
$300

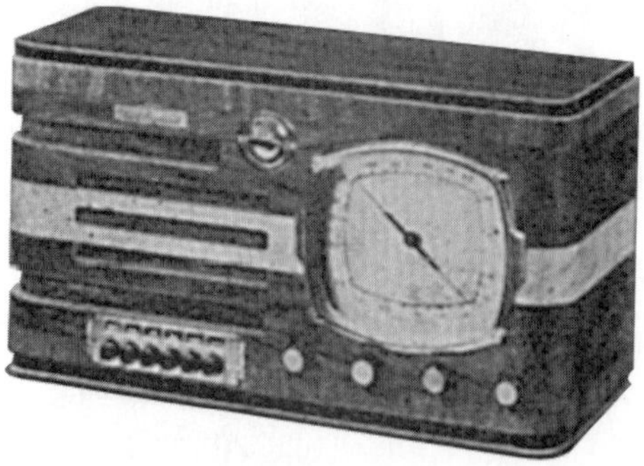

*Freed-Eisemann*

**88, c.1938**
7 tubes, 2 bands
$150

*Freed-Eisemann*

**FE97,98, c.1932**
FE97 - 8 tubes, 1 band, DC $175
FE98 - 8 tubes, 1 band, AC $250

Freed-Eisemann
**H357P, c.1936**
5 tubes, 2 bands
$125

Freed-Eisemann
**'Table', c.1933**
5 tubes, 1 band
$150

Freed-Eisemann
**'Treasure Chest', c.1933**
5 tubes, 1 bands
$350

***FROST-MINTON***
**'Bookset', c.1932**
4 tubes, 1 band
$400

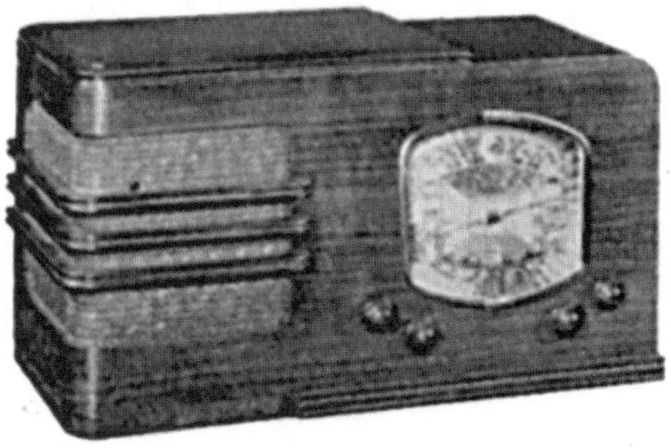

Garod
**8A4, c.1939**
8 tubes, 4 bands
$300

**GENERAL**
**521, c.1937**
5 tubes, 1 band
$190

**GENERAL**
**622, c.1937**
6 tubes, 1 band
$135

General
TELEVISION
**L47-1, c.1942**
5 tubes, 1 band, AC/DC
$150

General
TELEVISION
**L49, c.1942**
5 tubes, 1 bands
$225

General
TELEVISION

**L51, c.1942**
5 tubes, 1 band
$125

**GENERAL ELECTRIC**

**125,126,127, c.60**
5 tubes, 1 band
125 - pink $50, 126 - tan $25
127 - aqua $50

**GENERAL ELECTRIC**

**860 'Atomic', c.1959**
5 tubes, 1 band
$125

**GENERAL ELECTRIC**

**BX, c.1933**
metal, 4 tubes, 1 band
$125

**GENERAL ELECTRIC**

**C3C, c.1934**
metal, 3 tubes, 1 band
$175

**GENERAL ELECTRIC**

**C405,407, c.1960**
5 tubes, 1 band, clock/radio
C405-grey $50
C407-pink $75

**GENERAL ELECTRIC**

**C421,422, c.1960**
6 tubes, 1 band, clock/radio
C421-blue $125
C422-pink $110

**GENERAL ELECTRIC**

**C430,431, c.1960**
5 tubes, 1 band, clock/radio
C430-ivory $60
C431-pink $90

**GENERAL ELECTRIC**

**C435, c.1960**
5 tubes, 1 band, clock/radio
white $125

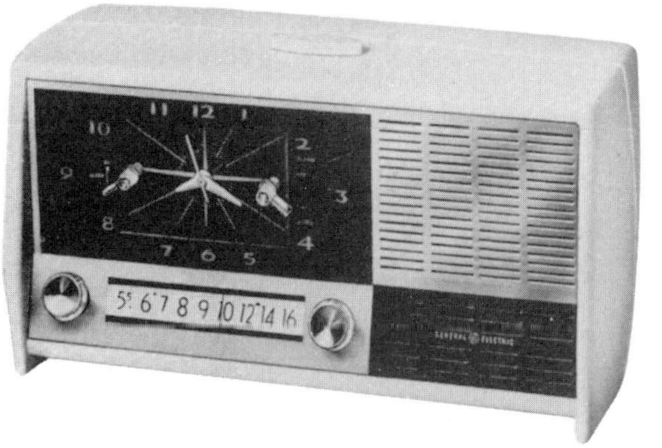

**GENERAL ELECTRIC**

**C440, c.1960**
5 tubes, 1 band, clock/radio
blue $75, black $50

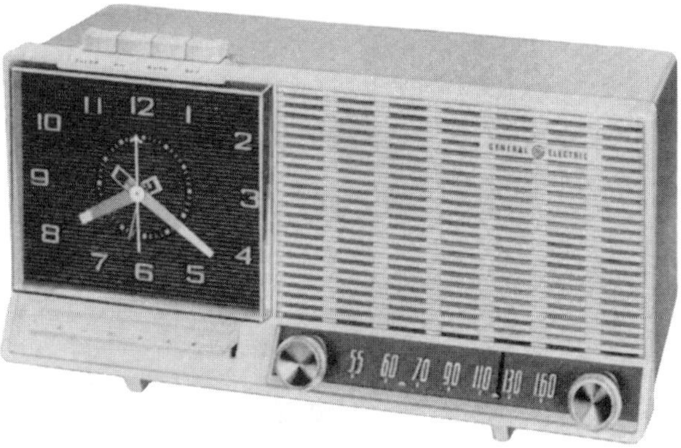

**GENERAL ELECTRIC**

**C450/451/452, c.1960**
5 tubes, 1 band, clock/radio
white $20, aqua $40, grey $25

**GENERAL ELECTRIC**

**CT455, c.1960**
transistor, 1 band, clock/radio
black & white $125

**GENERAL ELECTRIC**

**E6BX, c.1937**
6 tubes, 2 bands, DC
$65

**GENERAL ELECTRIC**

**G50, c.1939**
5 tubes, 1 band
$175

**GENERAL ELECTRIC**

**JB410, c.1941**
4 tubes, 1 band
$125

**GENERAL ELECTRIC**
**K51, c.1933**
5 tubes, 1 band
$275

**GENERAL ELECTRIC**
**K54, c.1933**
5 tubes, 1 band, radio/phono
$275

**GENERAL ELECTRIC**
**L50, c.1933**
repwood, 5 tubes, 1 band
$300

**GENERAL ELECTRIC**
**L610, c.1942**
6 tubes, 1 band
$40

**GENERAL ELECTRIC**
**L624, c.1942**
6 tubes, 2 bands
$60

**GENERAL ELECTRIC**
**LB424, c.1941**
4 tubes, 1 band, DC
$25

**GENERAL ELECTRIC**
**T130,131,132, c.1960**
5 tubes, 1 band
T130-pink $80, T131-blue $85
T132-grey $70

**GENERAL ELECTRIC**
**T135, c.1960**
9 tubes, AM/FM
white $45

**GENERAL ELECTRIC**
**T140, c.1960**
5 tubes, 1 band
T140-green $50, T141-tan $30

**GENERAL ELECTRIC**
**T145,146, c.1960**
transistorized, 1 band
T145-grey $75, T146-brown $60

**GENERAL ELECTRIC**
**T155,156,c.1960**
6 tubes, 1 band
T155-tan $35, T156-grey $40

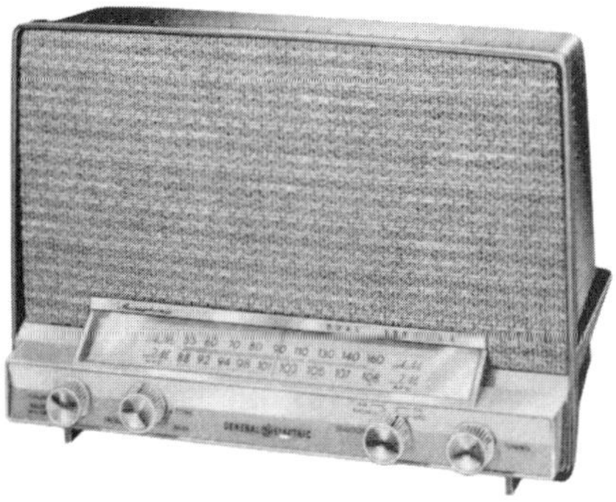

**GENERAL ELECTRIC**
**T120, c.1960**
9 tubes, AM/FM
tan $50

**GILBERT**
**70, c.1931**
6 tubes, 1 band
$600

*Gilfillan*
**590, c.1935**
6 tubes, 1 band
$275

*Gilfillan*
**831T, c.1937**
8 tubes, 3 bands
$225

GIMBELL
**'Treasure Chest', c.1931**
4 tubes, 1 band
$350

*Globe Trotter*
**3940, c.1939**
4,5 tubes, 1 band
brown $125, black $150
red $425, ivory $175

*Globe Trotter*
**3948, c.1939**
4,5 tubes, 1 band, radio/phono
$40

Globe Trotter
**3952, c.1939**
5 tubes, 1 band
brown $100, ivory $175

Globe Trotter
**3954, c.1939**
5 tubes, 1 band
brown $175, ivory $250, black $225

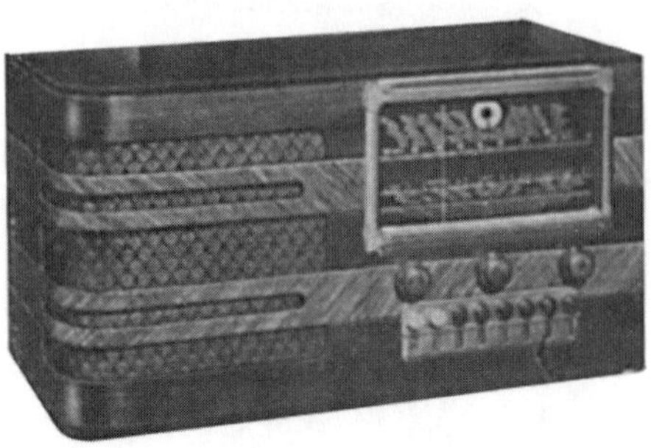

Globe Trotter
**A4201, c.1939**
7 tubes, 2 band
$125

Globe Trotter
**A4206, 1939**
8 tubes, 3 bands
$225

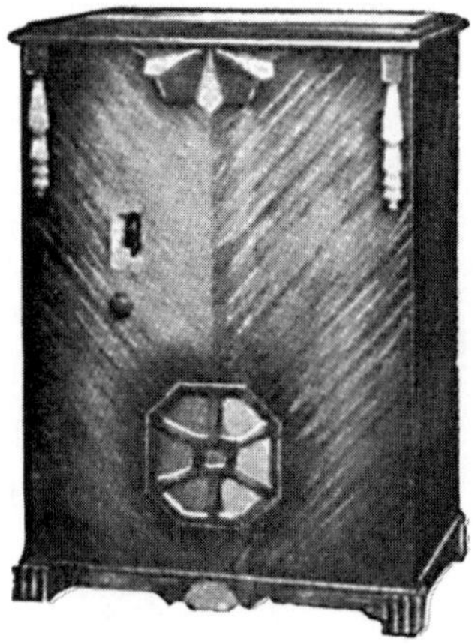

Gloritone
**27P, c.1931**
6 tubes, 1 band
$175

GOLDENTONE
**A3,A4, c.1942**
5 tubes, 1 band
A3-brown $75, A4-ivory $90

GOLDENTONE
**1A3,1A4, c.1942**
5 tubes, 1 band
1A3-brown $90, 1A4-ivory $100

GOLDENTONE
**1W11,1W12, c.1942**
5 tubes, 1 band
1W11-brown $65, 1W12-ivory $75

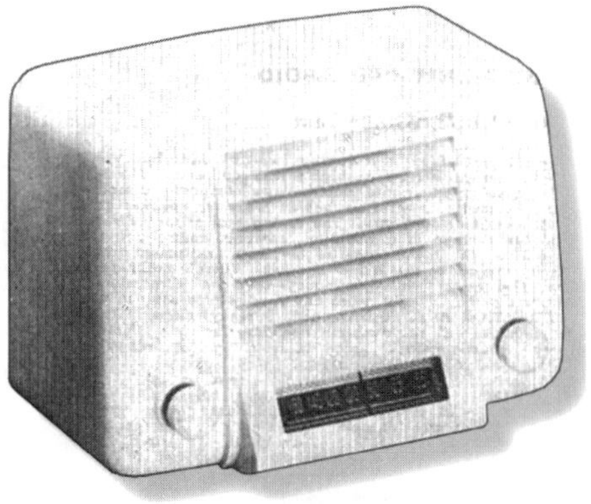

GOLDENTONE
**1W13,1W14, c.1942**
5 tubes, 1 band
1W13-brown $70, 1W14-ivory $80

**GOLDENTONE**

**1W15,1W16, c.1942**
6 tubes, 2 bands
1W15-brown $45, 1W16-ivory $50

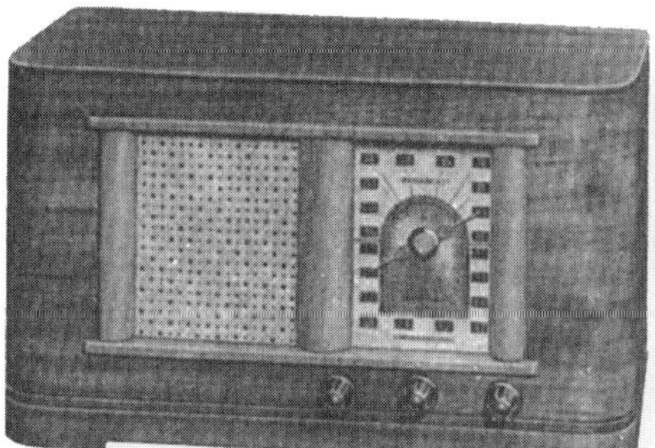

**GOLDENTONE**

**3A10, c.1942**
5 tubes, 2 bands
$45

**GOLDENTONE**

**3A11, c.1942**
6 tubes, 2 bands
$125

**GOLDENTONE**

**3A12, c.1942**
6 tubes, 2 bands
$110

**GOLDENTONE**

**3A8, c.1942**
5 tubes, 1 band
$50

**GOLDENTONE**

**3A9, c.1942**
5 tubes, 1 band
$165

**GOLDENTONE**

**3W10, c.1942**
7 tubes, 2 bands
$60

**GOLDENTONE**

**3W9, c.1942**
6 tubes, 2 bands
$45

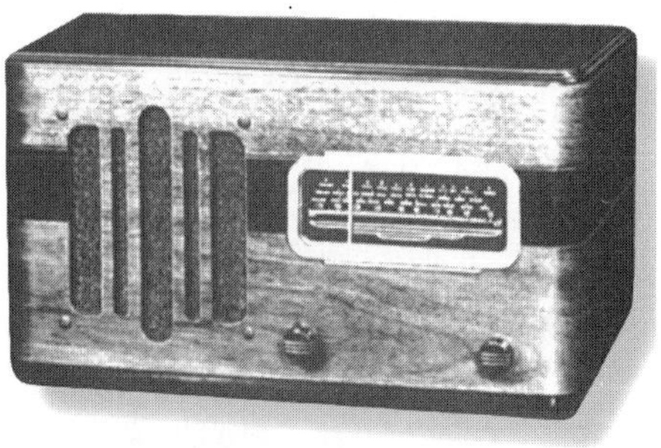

**GOLDENTONE**

**4A, c.1939**
4 tubes, 1 band
$20

**GOLDENTONE**
**4B1, c.1942**
75 tubes, 2 bands
$125

**GOLDENTONE**
**4X1, c.1942**
7 tubes, 2 bands
$50

**GOLDENTONE**
**4X2, c.1942**
9 tubes, 3 bands
$450

**GOLDENTONE**
**5Y, c.1939**
5 tubes, 2 bands
$165

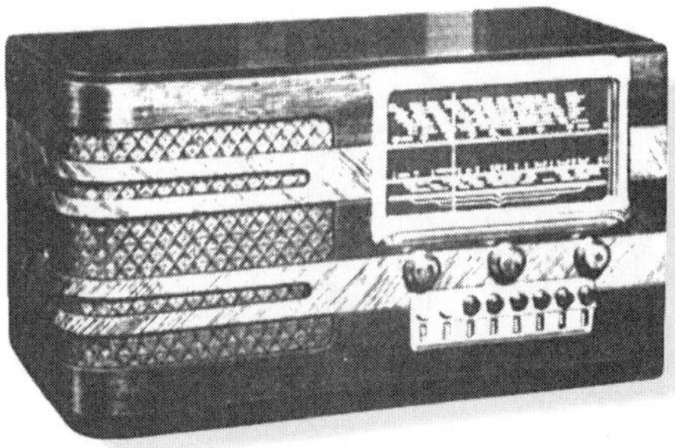

**GOLDENTONE**
**6C, c.1939**
6 tubes, 3 bands, DC
$60

**GOLDENTONE**
**6P1, c.1942**
6 tubes, 1 band
$25

**GOLDENTONE**
**6W10, c.1942**
4 tubes, 1 band, DC
$15

**GOLDENTONE**
**6W11, c.1942**
4 tubes, 1 band, DC
$20

**GOLDENTONE**
**6W12, c.1942**
5 tubes, 1 band
$25

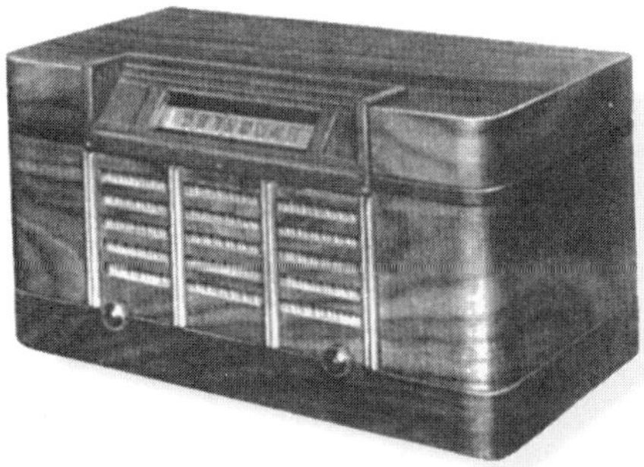

**GOLDENTONE**
**6W13, c.1942**
5 tubes, 1 band
$15

**GOLDENTONE**
**6W9, c.1942**
4 tubes, 1 band, DC
$90

**GOLDENTONE**
**7A6, c.1942**
4 tubes, 1 band, DC
$175

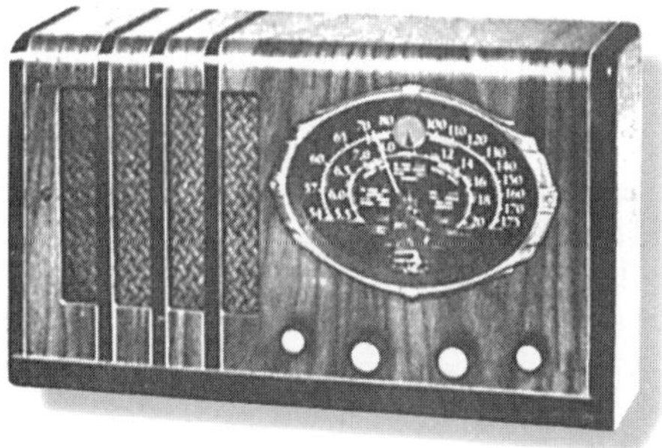

**GOLDENTONE**
**7D, c.1939**
7 tubes, 2 bands
$175

**GOLDENTONE**
**7S5, c.1942**
4 tubes, 1 band, DC
$150

**GOLDENTONE**
**9B3,c.1942**
5 tubes, 1 band
$25

**GOLDENTONE**
**9C13, c.1942**
5 tubes, 1 band, radio/phono
$35

**GOLDENTONE**
**9W3, c.1942**
5 tubes, 1 band, radio/phono
$30

**GOLDENTONE**
**A1, c.1940**
5 tubes, 1 band
$100

**GOLDENTONE**
**G1,G2, c.1940**
4 tubes, 1 band
G1-brown $125, G2-ivory $225

**GOLDENTONE**
**199B,199C, c.1939**
5 tubes, 1 band
199B-brown $250, 199C-ivory $450

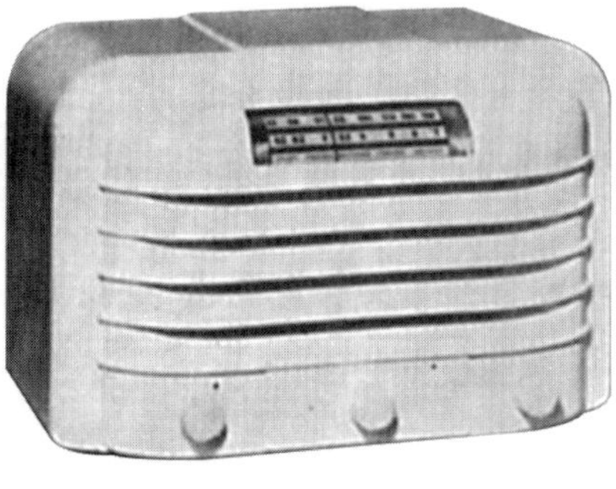

**GOLDENTONE**
**1A1, 1A2, c.1940**
6 tubes, 2 bands
1A1-brown $90, 1A2-ivory $125

**GOLDENTONE**
**1C1,1C2, c.1940**
5 tubes, 1 band
1C1-brown $90, 1C2-beetle $375

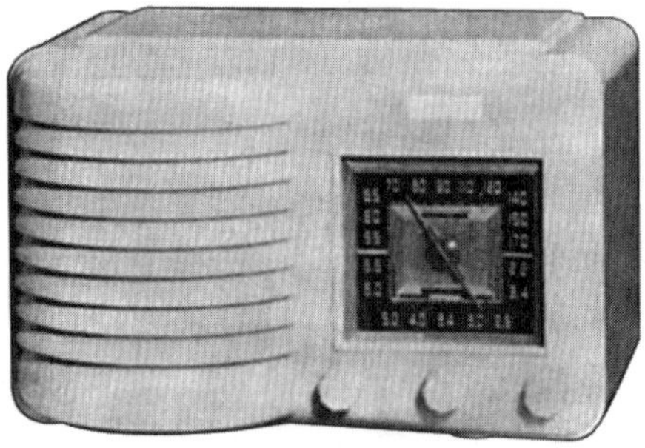

**GOLDENTONE**
**1G1,1G2, c.1940**
6 tubes, 2 bands
1G1-brown $175, 1G2-ivory $350

**GOLDENTONE**
**1W3,1W4, c.1940**
6 tubes, 2 bands
1W3-brown $150, 1W4-ivory $175

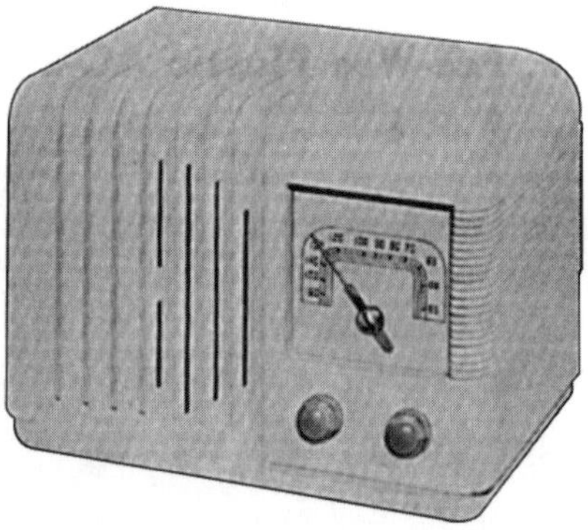

**GOLDENTONE**
**1W5,1W6, c.1940**
5 tubes, 1 band
1W5-brown $40, 1W6-ivory $50

**GOLDENTONE**
**208, c.1939**
5 tubes, 1 band
$75

**GOLDENTONE**
**208AP, c.1939**
5 tubes, 1 band, radio/phono
$75

**GOLDENTONE**
**209, c.1939**
8 tubes, 3 bands
$350

**GOLDENTONE**
**212, c.1939**
7 tubes, 2 bands
$120

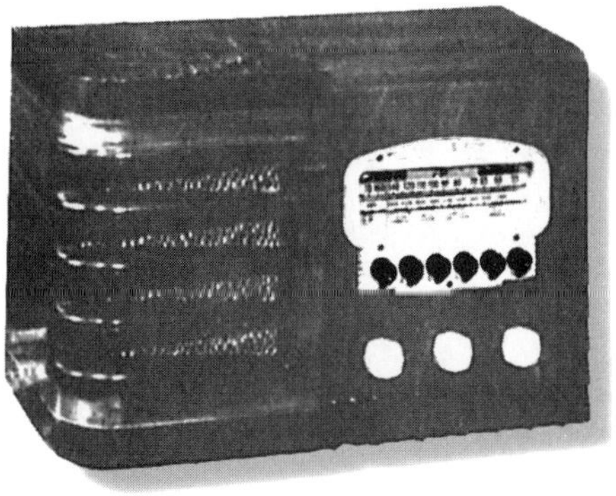

**GOLDENTONE**
**220, c.1939**
6 tubes, 2 bands
$250

**GOLDENTONE**
**221, c.1939**
6 tubes, 2 bands
$225

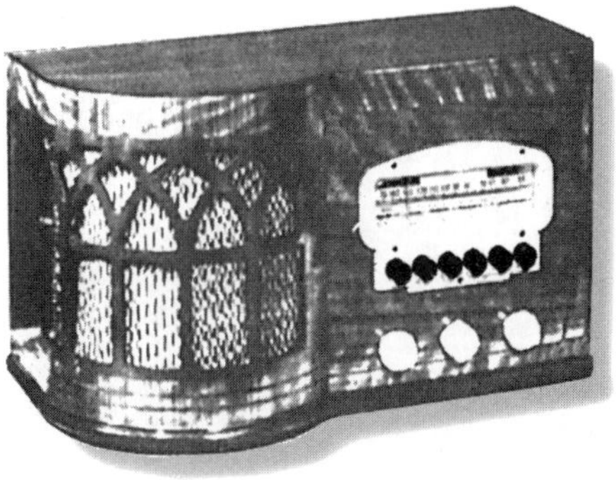

**GOLDENTONE**
**222, c.1939**
7 tubes, 2 bands
$350

**GOLDENTONE**
**223, c.1939**
7 tubes, 3 bands
$300

**GOLDENTONE**
**225, c.1939**
7 tubes, 3 bands
$250

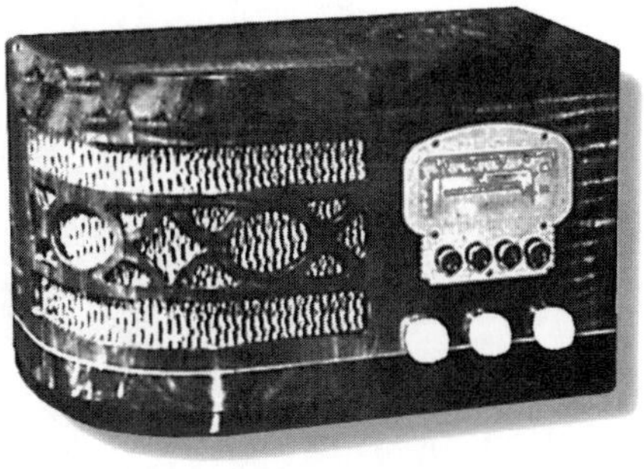

**GOLDENTONE**
**226, c.1939**
5 tubes, 2 bands
$300

**GOLDENTONE**
**231, c.1939**
9 tubes, 3 bands
$350

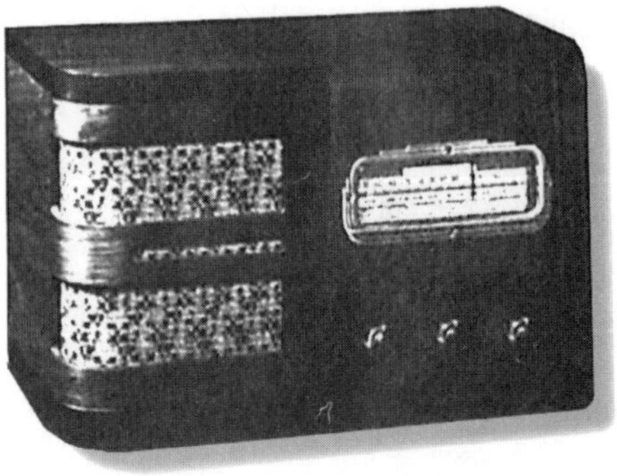

**GOLDENTONE**

**248, c.1939**
6 tubes, 2 bands
$125

**GOLDENTONE**

**249, c.1939**
6 tubes, 2 bands
$250

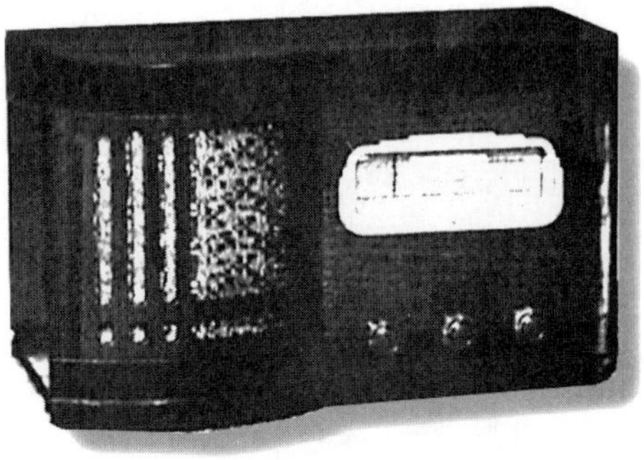

**GOLDENTONE**

**250, c.1939**
7 tubes, 2 bands
$225

**GOLDENTONE**

**251, c.1939**
7 tubes, 3 bands
$200

**GOLDENTONE**

**272, c.1939**
4 tubes, 1 bands
$150

**GOLDENTONE**

**2A1,2A2, c.1940**
6 tubes, 2 bands
2A1-brown $90
2A2-ivory $150

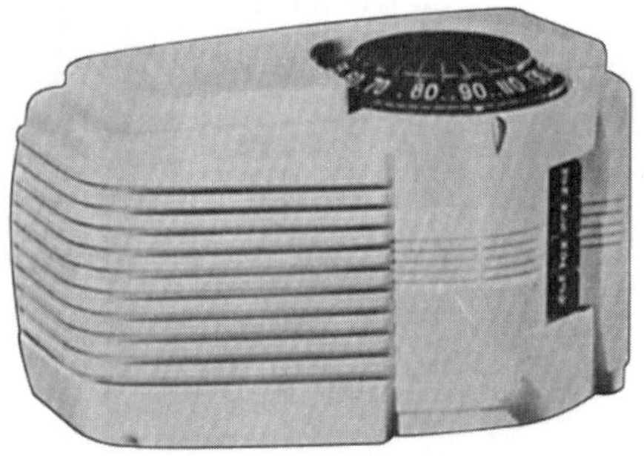

**GOLDENTONE**

**2S1,2S2, c.1940**
5 tubes, 1 band
2S1-brown $250
2S2-ivory $275

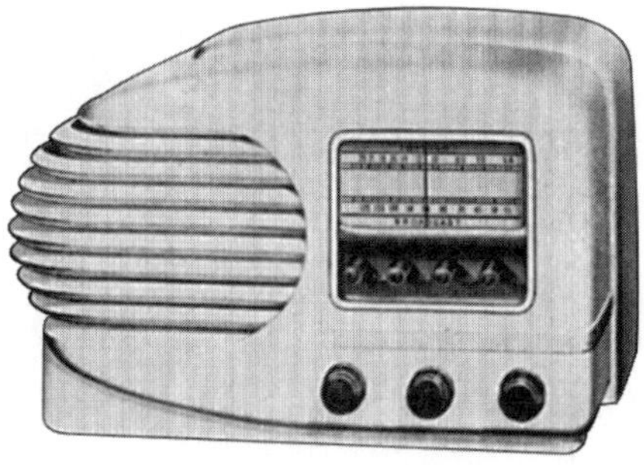

**GOLDENTONE**

**2W1,2W2, c.1940**
5 tubes, 1 band
2W1-brown $150
2W2-ivory $175

**GOLDENTONE**

**3A2, c.1940**
5 tubes, 1 band
$225

**GOLDENTONE**

**3G1, c.1940**
5 tubes, 1 band
$150

**GOLDENTONE**

**3G2, c.1940**
5 tubes, 1 band
$125

**GOLDENTONE**

**3G3, c.1940**
5 tubes, 1 band
$135

**GOLDENTONE**

**3T3, c.1940**
7 tubes, 2 bands
$70

**GOLDENTONE**

**3W3, c.1940**
7 tubes, 3 bands
$90

**GOLDENTONE**

**3W4,3W5, C.1940**
7,8 tubes, 3 bands
$75

**GOLDENTONE**

**3W6, C.1940**
5 tubes, 1 band
$150

**GOLDENTONE**

**400, c.1936**
5 tubes, 2 bands
$135

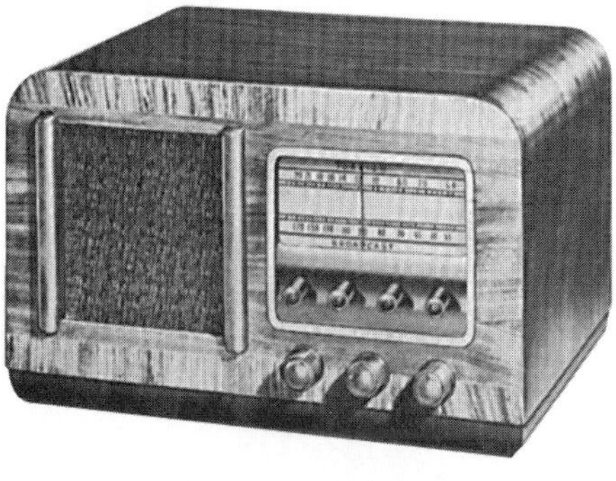

**GOLDENTONE**

**4W1, c.1940**
7 tubes, 2 bands
$75

**GOLDENTONE**
**4W2, c.1940**
7 tubes, 3 bands
$75

**GOLDENTONE**
**5C, c.1936**
5 tubes, 2 bands
$110

**GOLDENTONE**
**6C2, c.1936**
4 tubes, 1 band, DC
$25

**GOLDENTONE**
**6W2, c.1940**
4 tubes, 1 band, DC
$25

**GOLDENTONE**
**6W3, c.1940**
4 tubes, 1 band
$20

**GOLDENTONE**
**6W4, c.1940**
4 tubes, 1 band
$125

**GOLDENTONE**
**9A1, c.1940**
5 tubes, 1 band
$175

**GOLDENTONE**
**9A2, c.1940**
5 tubes, 1 band, radio/phono
$50

**GOLDENTONE**
**9A3, c.1940**
5 tubes, 1 band, radio/phono
$50

**GOLDENTONE**

**9A4 'Consolette', c.1940**
5 tubes, 1 band
$1500

**GOLDENTONE**

**9S1, c.1940**
5 tubes, 1 band, radio/phono
$125

**GOLDENTONE**

**9X2,9X4,9X5, c.1940**
phono remote
$15

**GOLDENTONE**

**'Baine', c.1939**
4 tubes, 1 band, DC
$110

**GOLDENTONE**

**D, c.1936**
5 tubes, 2 bands
$100

**GOLDENTONE**

**D7,DA7, c.1937**
D7 - 7 tubes, 2 bands $175
DA7 - 5 tubes, 2 bands, DC $75

**GOLDENTONE**

**G, c.1936**
6 tubes, 3 bands
$275

**GOLDENTONE**

**G9,GB9, c.1937**
G9 - 9 tubes, 3 bands $350
GB9 - 9 tubes, 3 bands, DC $200

**GOLDENTONE**

**GF6, c.1937**
6 tubes, 3 bands, DC
$90

**GOLDENTONE**

**IC12,13,c.1942**
5 tubes, 1 band
IC12-brown $45, IC13-ivory $50

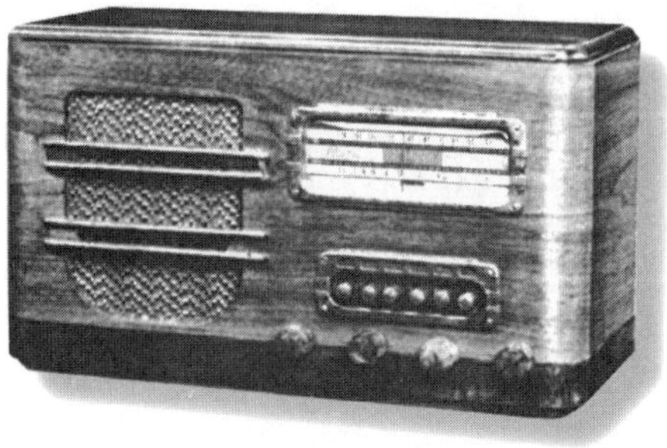

**GOLDENTONE**

**'Lunt',c.1939**
6 tubes,2 bands, DC
$20

**GOLDENTONE**

**NT, c.1940**
11 tube,3 bands, DC
$375

**GOLDENTONE**

**RJ5, c.1937**
5 tubes,2 bands
$175

**GOLDENTONE**

**SG9, c.1937**
9 tubes,3 bands
$450

**GOLDENTONE**

**NTR, c.1937**
11 tube,3 bands
$400

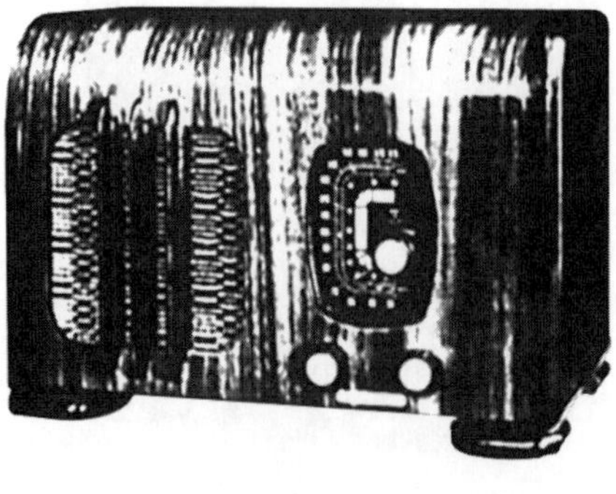

**GOLDENTONE**

**RJH, c.1939**
6 tubes,2 bands
$115

**GOLDENTONE**

**SD7, c.1937**
7 tubes,2 bands
$225

GOLDENTONE
**SG9, c.1937**
9 tubes,3 bands
$350

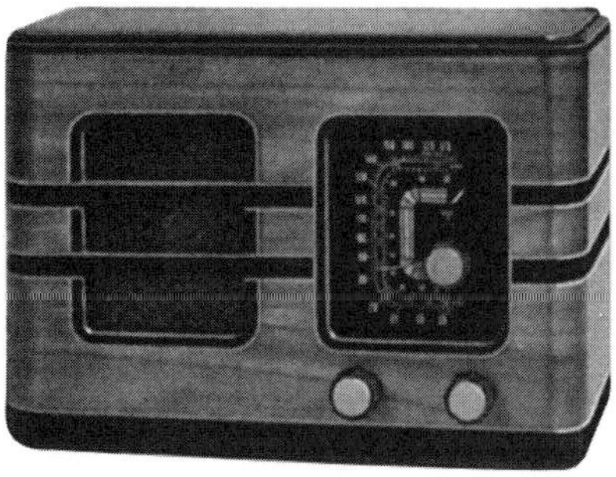

GOLDENTONE
**SRJ5, c.1937**
6 tubes,2 bands
$125

GOLDENTONE
**SZ4,Z4, c.1937**
4 tubes, 1 bands
$75

Granco
**T-160, c.1956**
8 tubes, FM Only
$50

Granco
**CR, c.1959**
8 tubes, AM/FM, clock/radio
grey $50, white $40

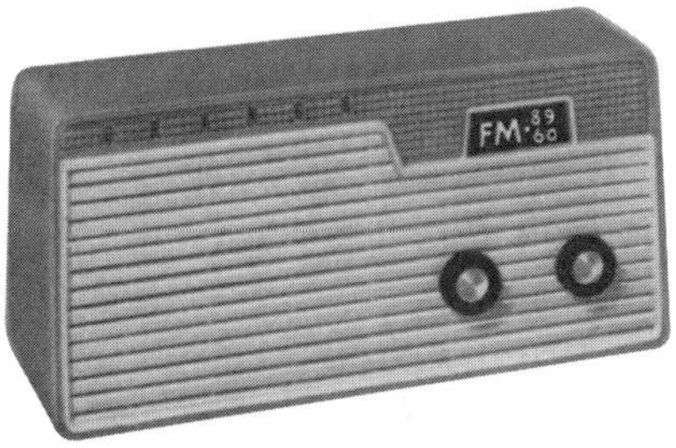

Granco
**FM, c.1959**
8 tubes, AM/FM
grey $110, aqua $135, pink $125

Granco
**FMAM, c.1959**
8 tubes, AM/FM
grey $110, aqua $135, maroon $125

GraybaR
**GT8, c.1932**
8 tubes, 1 band
$375

GraybaR
**GT7, c.1932**
7 tubes, 1 band
$225

**GRAYMARK**
**510, c.1956**
5 tubes, 1 band
$25

**CREBE**
**307L, c.1938**
7 tubes, 2 bands
$350

**CREBE**
**AHG9, c.1936**
9 tubes, 4 bands
$375

*Grunow*
**510, c.19DC**
5 tubes, 2 bands, DC
$750

*Grunow*
**550, c.1934**
5 tubes, 1 band
$225

*Grunow*
**566A, c.1938**
5 tubes, 2 bands
$250

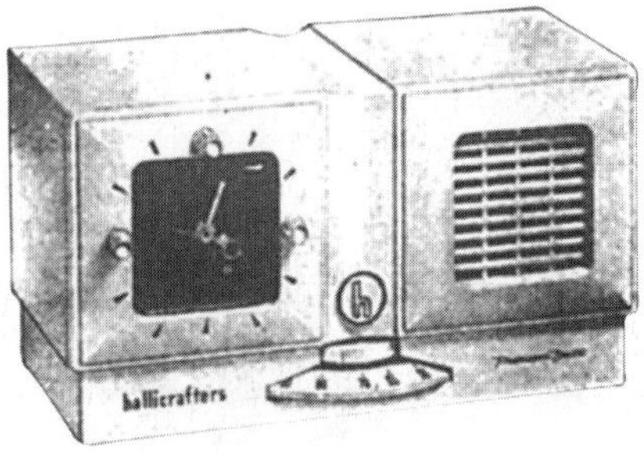


**hallicrafters**
**ATCL-11, c.1954**
5 tubes, 1 bands
$125

**Halson**
**624, c.1933**
4 tubes, 1 band
$125

**Halson**
**638, c.1933**
5 tubes, 1 bands
$225

**Halson**
**'Dwarf', c.1939**
5 tubes, 1 band, catalin
red $10000, green $8000, ivory $6500

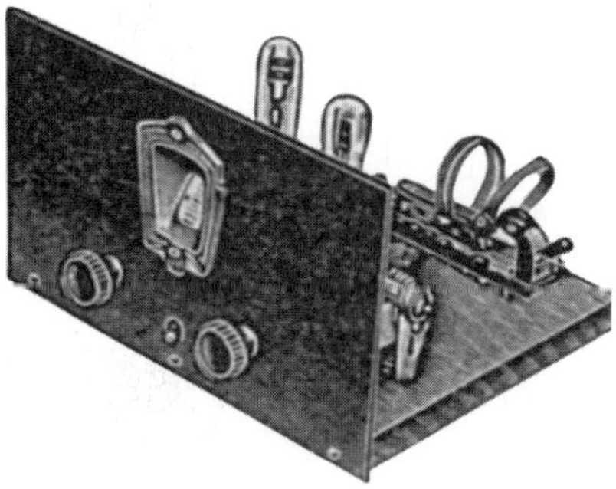

**Hammarlund**
**5-Tube SW Kit, c.1933**
5 tubes
$75

**HARRIS**
**701W, c.1937**
7 tubes, 2 bands, radio-phono
$225

**HOMCORD**
**11, c.1941**
11 tube, 3 bands, radio-phono-recorder
$325

**HOWARD**
20, c.1932
7 tubes, 1 band
$250

**HOWARD**
**245, c.1931**
5 tubes, 1 band, clock-radio
$225

**HOWARD**
**256, c.1937**
5 tubes, 2 bands
$95

**HOWARD**
**259T, c.1937**
7 tubes, 2 bands
$110

**HOWARD**
**266T, c.1937**
6 tubes, 3 bands
$125

**HOWARD**
**268T, c.1937**
7 tubes, 3 bands
$150

**HOWARD**
**302RT, c.1941**
7 tubes, 2 bands, radio-phono-recorder
$150

**HOWARD**
**307, c.1941**
5 tubes, 3 bands
$75

**HOWARD**
**468, c.1939**
7 tubes, 3 bands
$165

**HOWARD**
**780, c.1941**
8/10 tubes, 2 bands, bandspread
$95

**HOWARD**
**'Green Diamond', 1931**
6 tubes, 1 band
$275

**HOWARD**
**'Synchro-Dial', c.1933**
3 tubes, 1 band, remote tuner
$275

**HOWARD**
**X2, c.1934**
7 tubes, 1 band
$375

**HOWARD**

**SIC, c.1936**
6 tubes, 3 bands
$250

**Hudson-Ross**

**1G 'Three Little Pigs', c.1933**
4 tubes, 1 band
$3500

**Hudson**

**Jr. Juke 'Disneyland', c.1959**
3 tubes, phono remote
$225

**Hudson**

**R4Z, c.1936**
4 tubes, 1 band
$115

**Hudson**

**R5CT, c.1936**
5 tubes, 2 bands
$125

**Hudson**

**R5DKT, c.1936**
5 tubes, 2 bands
$170

**Hudson**

**R6GHT, c.1936**
6 tubes, 3 bands
$200

**Hudson**

**R8JT, c.1936**
8 tubes, 5 bands
$300

**Hudson**

**RA5, c.1936**
5 tubes, 1 band
$125

Hudson

**RCD, c.1936**
5 tubes, 1 band
$100

ICA

**'Conquerer', c.1932**
6 tubes, allwave
$150

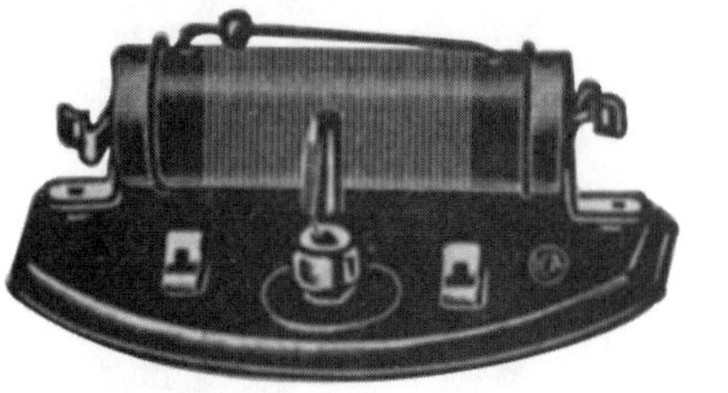

ICA

**Crystal Set, c.1940**
crystal receiver
$100

ICA

**'Mascot', c.1933**
5 tubes, 1 band
$250

IMPERIAL

**510, c.1941**
5 tubes, 1 band, radio-phono
$40

IMPERIAL

**'Courier', c.1933**
5 tubes, 1 band, clock-radio
$175

IMPERIAL

**'Mantel', c.1932**
4 tubes, 1 band
$150

Jackson Bell

**50 'Tulip Swirl', c.1932**
5 tubes, 1 band
$700

Jackson Bell

**62 'Lyre', c.1931**
6 tubes, 1 band
$425

Jackson Bell

**69 'Fan Grille', c.1931**
8 tubes, 1 band
$650

Jackson Bell

**82 'Tulip', c.1932**
6 tubes, 1 band
$750

JESSE FRENCH

**'Devon', c.1932**
7 tubes, 1 band
$500

JESSE FRENCH

**'Junior', c.1931**
5 tubes, 1 band
$600

JESSE FRENCH

**'Tudette', c.1932**
4 tubes, 1 band
$500

KADETTE

**K25/26/27/28, c.1938**
6 tubes, 1 band
catalin: K25 - alabaster $2000
K26 - red $7500, K27 - green $4000
K28 - blue $7500, transparent blue $10000

KADETTE

**41/43/44/47/48, c.1936**
4 tubes, 1 band
41 - brown $300, 43 - ivory $450
44 - red $750, 47 - black $400
48 - marbeled brown $750

KADETTE

**BW, c.1934**
5 tubes, 1 band
$400

KADETTE

**K10-16, c.1938**
5 tubes, 1 band
7 color combinations $900-1500
NOTE: Rarely found near perfect

INGRAHAM

## KADETTE

**L40, c.1940**
4 tubes, 1 band, clock-radio
$250

## KELLER-FULLER

**'Baby Midget', c.1931**
5 tubes, 1 band
$600

## KENNEDY
*The Royalty of Radio*

**1142 'Coronet', c.1932**
6 tubes, 1 band
$425

## KENNEDY
*The Royalty of Radio*

**'Globetrotter', c.1932**
2 tubes, shortwave converter
$450

## KNIGHT

**9505,9511,9515, c.1934**
9505 - 5 tubes, 2 bands, AC $200
9511 - 5 tubes, 2 bands, DC $100
9511 - 5 tubes, 1 bands, DC $100

## KNIGHT

**9521, c.1934**
4 tubes, 2 band
$200

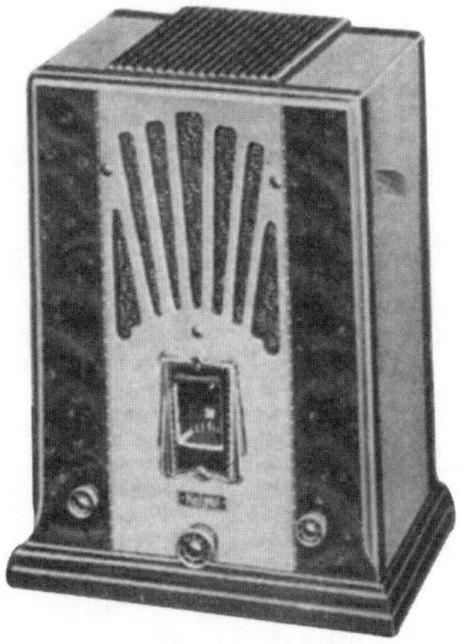

## KNIGHT

**9523, c.1934**
4 tubes, 2 bands
$350

## KNIGHT

**9527, c.1933**
5 tubes, 2 bands
$110

## KNIGHT

**9537, c.1933**
5 tubes, 1 band
$125

**KNIGHT**

**9555, c.1934**
5 tubes, 1 band
$175

**KNIGHT**

**9601,9602, c.1933**
4 tubes, 1 band
$150

**KNIGHT**

**9567, c.1934**
8 tubes, 1 band, DC
$160

**KNIGHT**

**9643,9663, c.1934**
9643 - 8 tubes, 1 band $300
9663 - 10 tubes, 2 bands $425

**KNIGHT**

**9687/85, c.1934**
9685 - 5 tubes, 1 band $160
9687 - 5 tubes, 2 bands $175

**KNIGHT**

**9701, c.1933**
4 tubes, 1 band
$350

**KNIGHT**

**9707, c.1933**
5 tubes, 1 band
$200

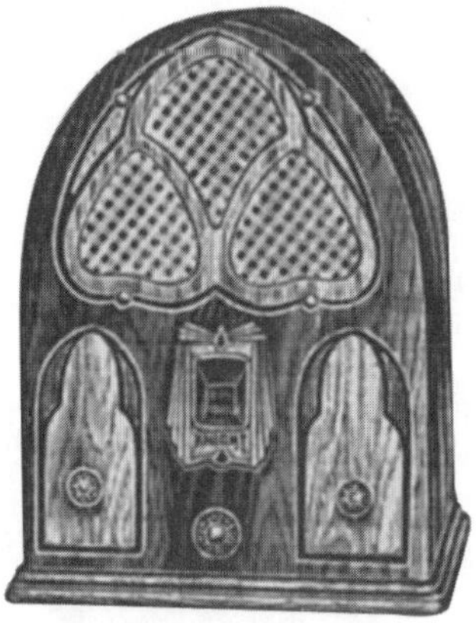

**KNIGHT**

**9717, c.1933**
4 tubes, shortwave
$250

## KNIGHT

**9721, 9727, c.1933**
9721 - 5 tubes, 2 bands, AC $225
9727 - 5 tubes, 2 bands, DC $125

## KNIGHT

**9751, c.1934**
5 tubes, 2 bands
$225

## KNIGHT

**9761, c.1934**
8 tubes, 4 bands
$500

## KNIGHT

**9824, c.1933**
8 tubes, 1 band, DC
$225

## KNIGHT

**9890 'Skyrider', c.1939**
4 tubes, shortwave
$150

## L.A. RADIO

**'Paramount', c.1932**
4 tubes, 1 band
$350

## Lafayette

**E22, 226, c.1933**
10 tubes, 3 bands
$375

## Lafayette

**L11, c.1933**
8 tubes, 2 bands, DC
$125

## Lafayette

**L18 'Dual Speaker', c.1933**
7 tubes, 1 band
$165

Lafayette

**L20 'Nomad', c.1933**
5 tubes, 1 band
$350

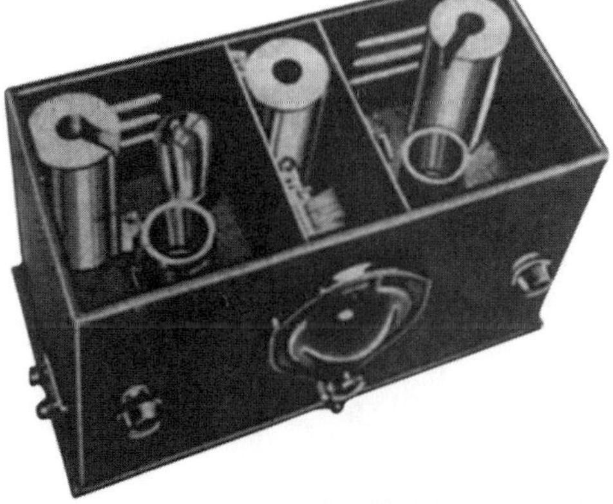

Lafayette

**M10 'Deluxe Shortwave', c.1933**
6 tubes
$350

Lafayette

**M31 'Minstrel, c.1933**
5 tubes, 1 band
$125

Lafayette

**M33,37,69, c.1933**
M33 - 6 tubes, 2 bands $150
M37 - 7 tubes, 1 band $150
M69 - 7 tubes, 3 bands $160

Lafayette

**M39,43,47, c.1932**
M39 - 8 tubes, 2 bands $175
M43 - 8 tubes, 3 bands $200
M47 - 10 tubes, 3 bands $250

Lafayette

**M53, c.1933**
7 tubes, 1 band, radio-phono
$175

Lafayette

**M57,59, c.1933**
M57 - 8 tubes, 3 bands $200
M59 - 10 tubes, 2 bands $250

Lafayette

**M71 'Thriller', c.1933**
5 tubes, 1 band
$200

Lafayette

**R71 'Little Treasure', c.1933**
4 tubes, 1 band
$110

Lafayette

**R73 'Souvenir', c.1933**
4 tubes, 1 band
$200

Lafayette

**S10 'Mayfair' c.1933**
4 tubes, 1 band
$400

Lafayette

**S91, c.1933**
4 tubes, 1 band, DC
$65

Lafayette

**S93, c.1933**
6 tubes, 1 band, DC
$100

Lafayette

**S99, c.1933**
6 tubes, 1 band
$200

Lafayette

**SGF6, c.1937**
6 tubes, 3 bands, DC
$75

Lafayette

**30, c.1931**
6 tubes, 1 band, DC
$165

Lafayette
**T20 'Trutest', c.1933**
4 tubes, 1 band
$115

Lafayette
**T30 'Pigmy', c.1933**
5 tubes, 1 band
$250

Legionairre
**'Pentode Midget' c.1932**
4 tubes, 1 band
$125

Lind's
**'Midget' c.1932**
4 tubes, 1 band
$225

M & H
**'Midget' c.1931**
5 tubes, 1 band
$350

LION
**101, c.1935**
1 tube, shortwave
$125

WURLITZER
LYRIC
**451E, c.1935**
4 tubes, 1 band
$500

WURLITZER
**LYRIC**

**470B, c.1935**
7 tubes, 2 bands
$750

WURLITZER
**LYRIC**

**'All American' c.1931**
6 tubes, 1 bands
$3000

WURLITZER
**LYRIC**

**B6L, c.1935**
6 tubes, 1 band, DC
$1500

WURLITZER
**LYRIC**

**J3, c.1931**
6 tubes, 1 band
$350

WURLITZER
**LYRIC**

**S50, c.1933**
5 tubes, 1 bands
$250

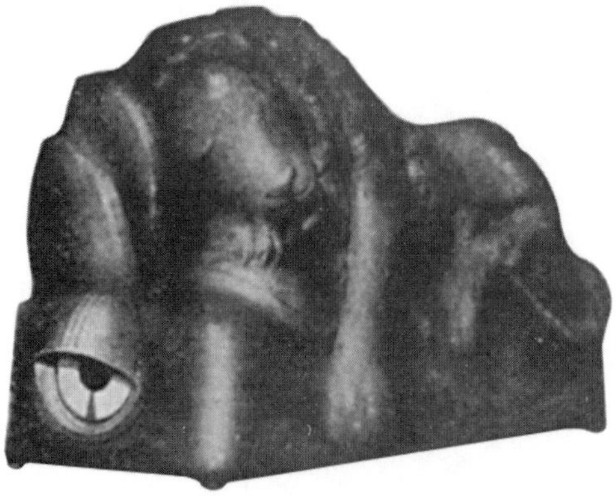

Magic Lion

**'Magic Lion'**
3 tube tuning eye attachment
$350

*Magnavox*

**31B, c.1940**
radio-phono
$60

*Magnavox*

**'Phonette', c.1937**
phono remote
$35

*Majestic* of CANADA

**761, c.1938**
6 tubes, 3 bands
$250

Majestic of CANADA

**770, c.1938**
7 tubes, 3 bands
$250

Majestic of CANADA

**4151, c.1941**
5 tubes, 1 band
$175

Majestic of CANADA

**4151AL, c.1941**
5 tubes, 1 band
$225

Majestic of CANADA

**4152, 4152X, c.1941**
5 tubes, 1 band
$110

Majestic of CANADA

**4153, c.1941**
5 tubes, 1 band
$135

Majestic of CANADA

**4154, c.1941**
5 tubes, 2 bands
$110

Majestic of CANADA

**4155, c.1941**
5 tubes, 6 bands
$125

Majestic

**11, c.1932**
3 tubes, shortwave converter
$225

Majestic

**130 'Scoop', c.1939**
4 tubes, 1 band, DC
$165

Majestic

**195, c.1933**
5 tubes, 1 band
$275

Majestic

**1A50A, c.1940**
5 tubes, 1 band
$90

Majestic

**1A59, c.1939**
5 tubes, 1 band
$110

Majestic

**201 'Sheffield', c.1932**
7 tubes, 1 band
$225

Majestic

**250W 'Zephyr', c.1940**
6 tubes, 1 band
$175

Majestic

**259EB, c.1939**
6 tubes, 2 bands
$125

Majestic

**4170A, c.1941**
7 tubes, 6 bands
$165

Majestic

**4482, c.1940**
5 tubes, 1 band
$50

Majestic

**4PWO, c.1939**
4 tubes, phono only
$75

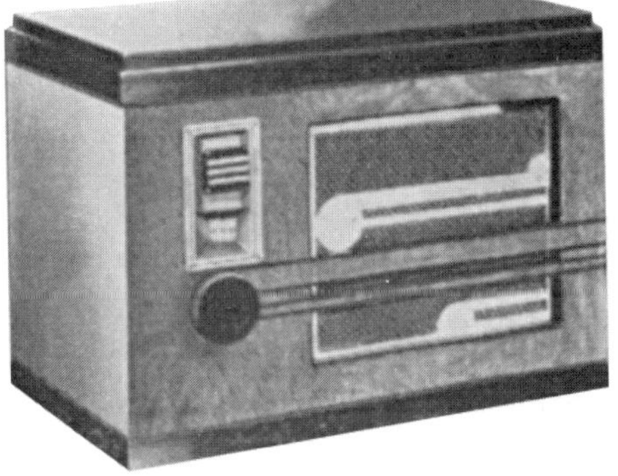

Majestic

**55 'Duette', c.1934**
5 tubes, 1 band
$950

Majestic

**56 'Ardmore', c.1932**
6 tubes, 1 band
$300

Majestic

**599 'Radiograph', c.1934**
5 tubes, 1 band, radio-phono
$375

Majestic

**42 'Deluxe Compact', c.1932**
8 tubes, 1 band
with stand -$250
without stand - $175

Majestic

**5A410,W 'Zephyr', c.1940**
5 tubes, 1 band
5A410 - walnut- $175
5A410W- ivory- $190

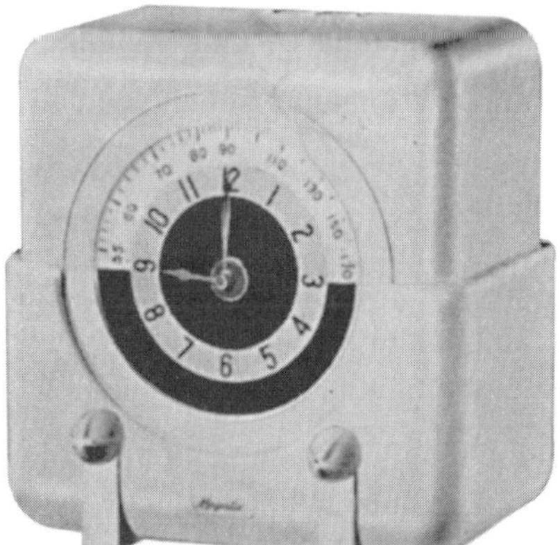

Majestic

**5T,5TO, c.1940**
5 tubes, 1 band
5T - ivory $400, 5TO - beetle $750

Majestic

**62A, c.1939**
6 tubes, 2 bands
$250

Majestic

**7P420B, c.1941**
7 tubes, 1 band
$75

*Majestic*
**7YR752, c.1947**
7 tubes, 1 band, radio-phono
$40

*Majestic*
**8FM744, c.1947**
8 tubes, AM/FM
$50

*Majestic*
**P1A50, c.1940**
6 tubes, 1 band, radio-phono
$75

*Majestic*
**P1A59, c.1939**
5 tubes, 1 band, radio-phono
$90

*Majestic*
**Pacemaker, c.1934**
6 tubes, 1 band
$850

*Majestic*
**T101L-A, c.1941**
5 tubes, 1 band
$35

*Majestic*
**TR321A, c.1941**
7 tubes, 1 band, radio-phono
$40

*Mantola*
**R402/3, c.1938**
5 tubes, 1 band
R402 - black $275, R403 - ivory $375

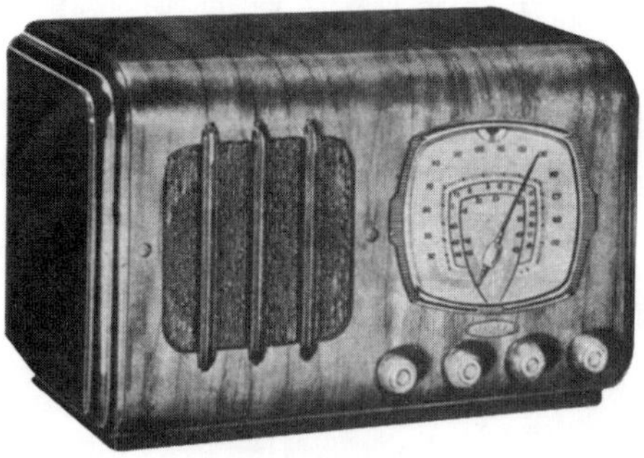

*Mantola*
**R404, c.1938**
6 tubes, 2 bands
$75

**MARCONI**
**108, c.1938**
6 tubes, 2 bands, DC
$125

**MARCONI**
**110, c.1938**
4 tubes, 1 band
$500

**MARCONI**
**112, c.1938**
5 tubes, 2 bands
$350

**MARCONI**
**113, c.1938**
5 tubes, 3 bands
$125

**MARCONI**
**115, c.1938**
5 tubes, 3 bands
$225

**MARCONI**
**153, c.1938**
5 tubes, 1 band
$75

**MARCONI**
**154, c.1938**
5 tubes, 1 band
$60

**MARCONI**
**156, c.38**
5 tubes, 1 band
$75

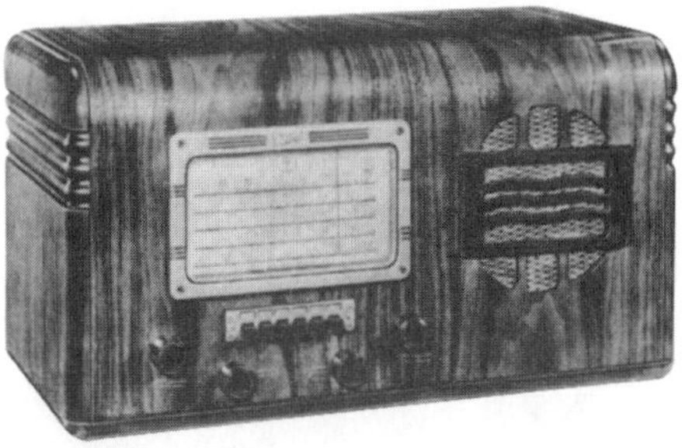

**MARCONI**
**157, c.1939**
5 tubes, 5 bands
$90

**MARCONI**
**159, c.1939**
7 tubes, 3 bands
$115

**MARCONI**
**164, c.1939**
5 tubes, 1 band
$30

**MARCONI**
**166, c.1939**
5 tubes, 2 bands
$60

**MARCONI**
**167, c.1939**
5 tubes, 5 bands
$110

**MARCONI**
**169, c.1940**
7 tubes, 3 bands
$135

**MARCONI**
**180, c.1940**
5 tubes, 1 band
$90

**MARCONI**
**180A, c.1940**
5 tubes, 1 band
$125

**MARCONI**
**181, c.1940**
5 tubes, 1 band
$110

**MARCONI**
**181A, c.1940**
5 tubes, 1 band
$90

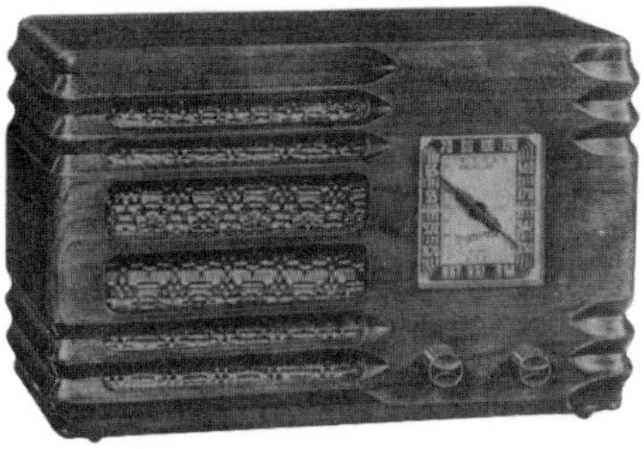

**MARCONI**
**182, c.1940**
5 tubes, 2 bands
$125

**MARCONI**
**184, c.1940**
5 tubes, 2 bands
$125

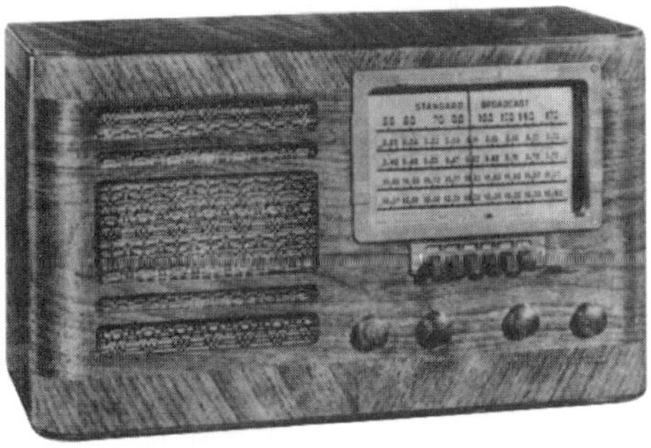

**MARCONI**
**185, c.1940**
6 tubes, 5 bands
$110

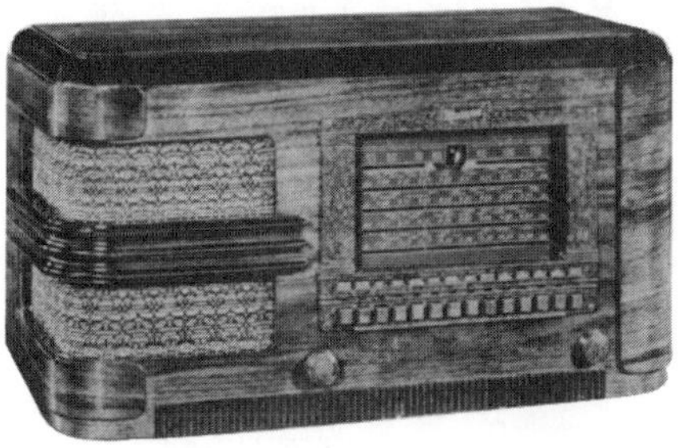

**MARCONI**
**187, c.1940**
6 tubes, 5 bands
$135

**MARCONI**
**193, c.1940**
4 tubes, 1 band/DC
$50

**MARCONI**
**194, c.1940**
4 tubes, 1 band, DC
$20

**MARCONI**
**196, c.1940**
5 tubes, 2 bands, DC
$65

**MARCONI**
**198, c.1940**
6 tubes, 5 bands, DC
$25

**MARCONI**
**200A, c.1940**
5 tubes, 1 band, radio-phono
$125

**MARCONI**
**200C, c.1940**
5 tubes, 1 band, radio-phono
$125

**MARCONI**
**201, c.1940**
6 tubes, 2 bands
$75

**MARCONI**
**201A, c.1940**
6 tubes, 2 bands
$175

**MARCONI**
**202, c.1940**
6 tubes, 2 bands
$75

**MARCONI**
**206, c.1940**
7 tubes, 6 bands
$135

**MARCONI**
**212, c.1941**
4 tubes, 1 band, DC
$25

**MARCONI**
**213, c.1941**
5 tubes, 1 band, DC
$45

**MARCONI**
**214, c.1941**
5 tubes, 2 bands, DC
$30

**MARCONI**
**270, c.1941**
7 tubes, 3 bands, DC
$125

**MASCOT**

**'Midget', c.1931**
7 tubes, 1 band
$375

*Master*

**'Viking', c.1935**
4 tubes, 1 band
$75

McCORMICK

**Mantle Clock, c.1931**
6 tubes, 1 band, clock-radio
$325

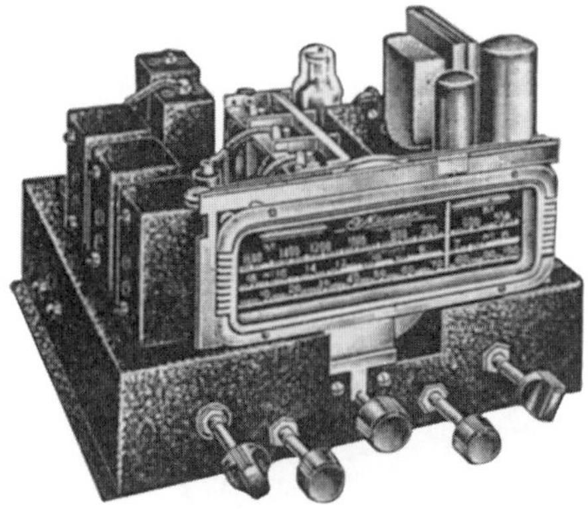

*Meissner*

**'Battery Super', c.1940**
6 tubes, 2 bands, DC
$60

*Meissner*

**'Combination Super', c.1940**
8 tubes, 4 bands
$225

*Meissner*

**'Custom Super', c.1940**
12 tubes, 5 bands
$350

*Meissner*

**'DeJong 8', c.1936**
8 tubes, 4 bands
$225

*Meissner*

**'Iron Core TRF', c.1940**
6 tubes, 1 band
$75

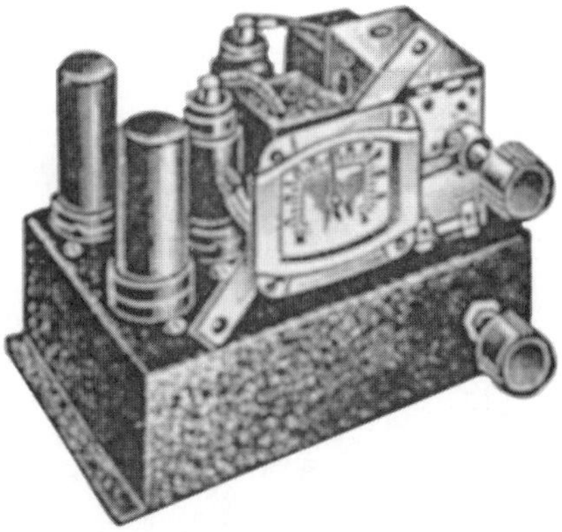

*Meissner*

**'Junior 4', c.1940**
4 tubes, 1 band, AC/DC
$50

Meissner

**'Junior', c.1940**
5 tubes, 1 band
$60

Meissner

**'Utility Super', c.1940**
7 tubes, 2 bands
$125

**MERCURY**

**5-32T 'Alladin', c.1936**
5 tubes, 1 band, DC
$75

**MERCURY**

**7-32T 'Beaumont', c.1936**
7 tubes, 1 band, DC
$75

**MERCURY**

**9-32T 'Berkley', c.1936**
9 tubes, 1 band, DC
$110

*Meridian*

**558, c.1947**
5 tubes, 1 band, radio-phono
$40

*Meridian*

**579, c.1947**
6 tubes, 1 band, AC/DC
$50

**MIDWEST**

**BS-9, c.1950**
9 tubes, 3 bands, bandspread
$90

**MIDWEST**

**HH7, c.1936**
7 tubes, 5 bands
$750

## MIDWEST

**JM-8, c.1950**
8 tubes, 5 bands
$150

## MIDWEST

**JV-5, c.1950**
5 tubes, 1 band, AC/DC
$75

## MIDWEST

**P8, c.1947**
6 tubes, 2 bands, AC/DC
$60

## MIDWEST

**Q11,16, c.1942**
Q11 - 11 tubes, 9 bands, bandspread - $300
Q16 - 16 tubes, 9 bands, bandspread - $425

## MIDWEST

**RM16 & 'Soundcabinet' Extension Speaker, c.1934**
RM16 -16 tubes, 5 bands - $900
Soundcabinet extension speaker - $600

**MIDWEST**
**R8, c.1942**
8 tubes, 9 bands, bandspread
$275

**MIDWEST**
**T5, c.1942**
5 tubes, 1 band
$175

**MIDWEST**
**T6, c.1942**
6 tubes, 9 bands, bandspread
$350

**MIDWEST**
**TM8, c.1947**
8 tubes, 5 bands
$200

**MISSION**
**387B 'Tempo', c.1938**
6 tubes, 2 bands
$275

**MISSION**
**Converter, c.1932**
shortwave converter
$125

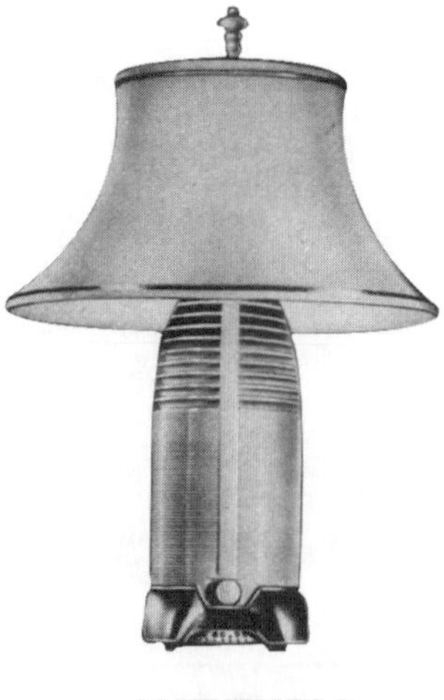

**MITCHELL**
**M-KC90 'Lumitone', c.1941**
5 tubes, 1 band
$275 with orig. shade/$225 without

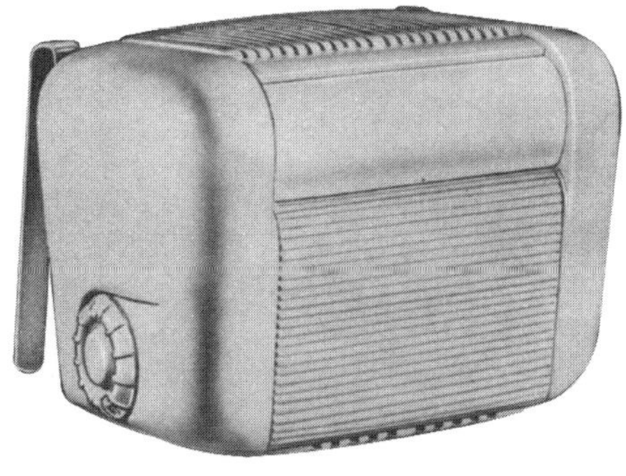

**MITCHELL**

**M-LB127,133 'Lullabye', c.1941**
5 tubes, 1 band
M-LB127 - ivory $125
M-LB133 - brown $110

**MITCHELL**

**M-TSG 'Globe', c.1941**
5 tubes, 1 band
$750

**Mohawk**

**C 'Blackhawk', c.1930**
7 tubes, 1 band
$375

**MONARCH**

**'Monarch', c.1938**
5 tubes, 1 band
$75

**Motorola**

**22A, c.1941**
2 tubes, phono remote
$25

**Motorola**

**22B, c.1042**
2 tubes, phono remote
$25

**Motorola**

**23RC, c.1941**
2 tubes, phono remote
$25

**Motorola**

**23RCW, c.1941**
2 tubes, phono remote
$30

**Motorola**

**40-60W, c.19-40**
6 tubes, 2 bands
$125

Motorola
**40B1, c.1941**
4 tubes, 1 band, DC
$75

Motorola
**40BW, c.1941**
4 tubes, 1 band, DC
$50

Motorola
**40B14, c.1941**
4 tubes, 1 band, DC
$50

Motorola
**45B12, c.1941**
4 tubes, 1 band, DC
$20

Motorola
**50W, c.1941**
5 tubes, 1 band
$30

Motorola
**50X1,2, c.1941**
5 tubes, 1 band
50X1 - brown $90
50X2 - ivory $110

Motorola
**50XC1/2/3/4, c.1940**
5 tubes, 1 band
catalin: 50XC1 - red $6000, 50XC2 - blue $7000
50XC3 - alabaster $5000, 50XC4 - green $7000

Motorola
**50XW, c.1941**
5 tubes, 1 band
$75

Motorola
**51F12, c.1942**
5 tubes, 1 band, radio-phono
$50

Motorola
**51X19, c.1942**
5 tubes, 1 band
$125

Motorola
**52T1, c.1941**
5 tubes, 2 bands
$75

Motorola
**52XAH1, c.1941**
5 tubes, 2 bands
$175

Motorola
**55F11, c.46**
5 tubes, 1 band, radio-phono
$30

Motorola
**55X13, c.1946**
5 tubes, 1 band
$25

Motorola
**56X1, c.1941**
5 tubes, 1 band
$110

Motorola
**56XA1, c.1941**
5 tubes, 1 band
$125

Motorola
**56XAW, c.1941**
5 tubes, 1 band
$125

Motorola
**58A11,12, c.1949**
5 tubes, 1 band
58AQ11 - brown $50
58AQ11 - ivory $60

**Motorola**

**58F1, c.1941**
5 tubes, 1 band, radio-phono
$50

**Motorola**

**58FRC, c.1941**
5 tubes, 1 band, radio-phono
$50

**Motorola**

**58R11-13, c.1949**
5 tubes, 1 band
brown $90, maroon $150, green $150
yellow $175, tan $125, ivory $100

**Motorola**

**58X11,12, c.1949**
5 tubes, 1 band
58X11 - brown $40
58X12 - ivory $50

**Motorola**

**60W, c.1941**
6 tubes, 2 bands
$75

**Motorola**

**60X1,2, c.1941**
6 tubes, 1 band
60X1 - brown $110
60X2 - ivory $125

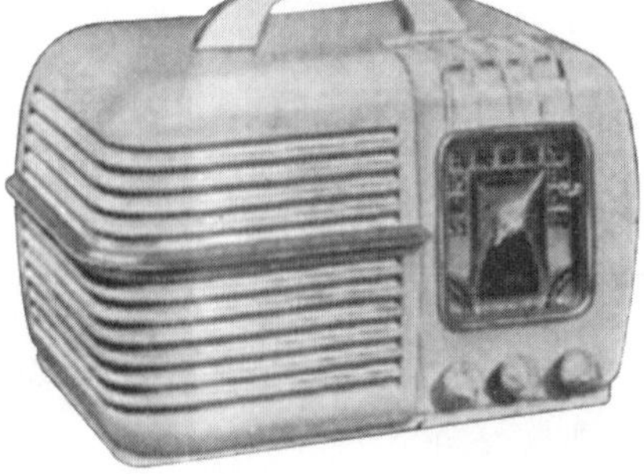

**Motorola**

**60XA1,2, c.1941**
6 tubes, 1 band
60XA1 - brown $140
60XA2 - ivory $160

**Motorola**

**60XW, c.1941**
6 tubes, 1 band
$75

**Motorola**

**61X17, c.1942**
6 tubes, 1 band, AC/DC
$45

**Motorola**

**62T1, c.1941**
6 tubes, 2 bands
$75

**Motorola**

**65F11, c.1946**
6 tubes, 1 band, radio-phono
$30

**Motorola**

**65T22, c.1946**
6 tubes, 2 bands
$75

**Motorola**

**65X11,12, c.1946**
6 tubes, 1 band
65X11 - brown $40
65X12 - ivory $50

**Motorola**

**65X14,14B, c.1946**
6 tubes, 1 band
65X14 - walnut $40
65X14B - blonde $50

**Motorola**

**M68F11, c.1949**
6 tubes, 1 band, radio-phono
$65

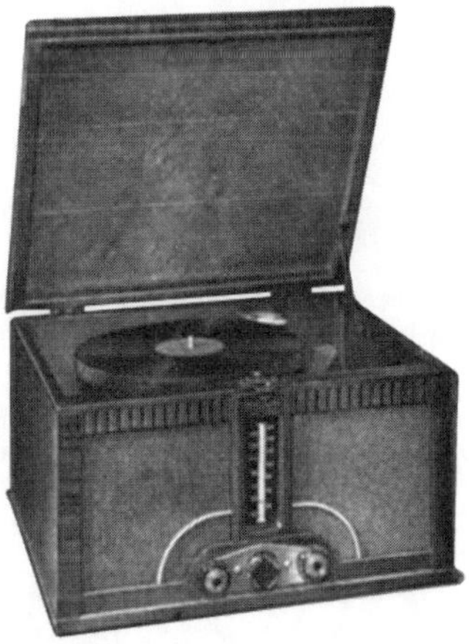

**Motorola**

**68F12, c.1949**
6 tubes, 1 band, radio-phono
$40

**Motorola**

**68T11, c.1949**
6 tubes, 1 band
$40

**Motorola**

**68T12, c.1949**
6 tubes, 1 band
$25

Motorola
**68X11,12, c.1949**
6 tubes, 1 band
68X11 - maroon $125
68X12 - brown $100

Motorola
**75T31, c.1946**
7 tubes, AM/FM
$80

Motorola
**77XM22, c.1949**
5 tubes, 1 band
$125

Motorola
**7T15-5K1, c.1947**
5 tubes, 1 band
$25

Motorola
**AR1, c.1942**
phono remote
$25

Motorola
**FM82, c.1942**
8 tubes, FM converter
$50

Motorola
**B150 'Bicycle Radio', c.1941**
3 tubes, 1 band, DC
$125

Motorola
**RW1, c.1941**
5 tubes, remote phono recorder
$75

Motorola
**WR3, c.1942**
phono remote
$25

Motorola
**WR4, c.1942**
phono remote
$30

Motorola
**WR6, c.1946**
phono remote
$25

Motorola
**WR7, c.1946**
phono remote
$30

Music Master
**518, c.1937**
5 tubes, 1 band
$60

Music Master
**521, c.1937**
5 tubes, 1 band
$150

Music Master
**620, c.1937**
6 tubes, 2 bands
$125

Music Master
**622, c.1937**
6 tubes, 1 band
$75

Music Master
**700, c.1938**
5 tubes, 1 band
$50

**Music Master**

**C4, c.1937**
4 tubes, 1 band
$75

**Music Master**

**D56,562, c.1937**
D56 - 5 tubes, 1 band $100
D562 - 5 tubes, 1 band $110

**Music Master**

**PETITE, c.1931**
5 tubes, 1 band
$250

*Mutual*

**3040, c.1935**
4 tubes, 1 band
$100

*Mutual*

**3041, c.1935**
4 tubes, 1 band
$110

*Mutual*

**3050, c.1935**
5 tubes, 1 band
$225

*Mutual*

**3051, c.1935**
5 tubes, 2 bands
$600

*Mutual*

**3052, c.1935**
5 tubes, 2 bands
$135

*Mutual*

**3053, c.1935**
5 tubes, 2 bands
$175

Mutual

**3054, c.1935**
6 tubes, 2 bands
$250

Mutual

**3056, c.1935**
5 tubes, 2 bands
$375

Mutual

**3058, c.1935**
5 tubes, 2 bands, DC
$125

Mutual

**3063, c.1935**
6 tubes, 3 bands
$135

Mutual

**3066, c.1935**
6 tubes, 1 band, DC
$100

Mutual

**3071, c.1935**
7 tubes, 2 bands, DC
$125

Mutual

**3080, c.1935**
8 tubes, 5 bands
$350

Mutual

**A682, A3282, c.1936**
A682 - 6 tubes, 1 band, DC $100
A3282 - 6 tubes, 1 band, DC $110

Mutual

**A6322, c.1936**
6 tubes, 3 bands
$250

Mutual
**B40, c.1935**
4 tubes, 1 band
$125

Mutual
**B45, c.1935**
4 tubes, 1 band
$225

Mutual
**B50, c.1935**
5 tubes, 1 band
$175

Mutual
**B52, c.1935**
5 tubes, 2 bands
$200

Mutual
**B55, c.1935**
5 tubes, 1 band
$135

Mutual
**B60, c.1935**
5 tubes, 2 bands
$175

Mutual
**B101, c.1935**
5 tubes, 2 bands
$75

Mutual
**B105, c.1935**
6 tubes, 3 bands
$175

Mutual
**B110, c.1935**
6 tubes, 3 bands
$275

*Mutual*

**B125, c.1935**
6 tubes, 3 bands
$200

*Mutual*

**B170, c.1935**
7 tubes, 3 bands
$150

*Mutual*

**B600, c.1935**
6 tubes, 3 bands, DC
$110

*Mutual*

**B600, c.1935**
6 tubes, 3 bands, DC
$110

*Mutual*

**N406, c.1936**
4 tubes, 1 band
$125

*Mutual*

**N407, c.1936**
4 tubes, 1 band
$225

*Mutual*

**N409, c.1936**
4 tubes, 1 band
$150

*Mutual*

**N409B, c.1936**
4 tubes, 1 band
$125

*Mutual*

**N415, c.1936**
4 tubes, 2 bands, TRF
$175

Mutual
**N508, c.1936**
5 tubes, 2 bands
$150

Mutual
**N521, c.1936**
5 tubes, 2 bands
$175

Mutual
**N523, c.1936**
5 tubes, 3 bands
$150

Mutual
**N551, c.1936**
5 tubes, 2 bands
$75

Mutual
**N601, c.1936**
6 tubes, 3 bands
$200

MUZAK
**Table Model, c.1936**

$225

National Union
**W2, c.1942**
5 tubes, 1 band
$125

Northern Electric
**320 'Glamis', c.1937**
3 tubes, 1 band
$150

Northern Electric
**414 'Little Trooper', c.1937**
4 tubes, 1 band
$100

Northern Electric

**420, c.1938**
4 tubes, 1 band
$125

Northern Electric

**518, c.1937**
5 tubes, 2 bands
$135

Northern Electric

**522 'Dublin', c.1938**
5 tubes, 2 bands, DC
$75

Northern Electric

**524, c.1937**
5 tubes, 1 band
$110

Northern Electric

**620, c.1937**
6 tubes, 3 bands
$135

Northern Electric

**622, c.1937**
6 tubes, 3 bands, DC
$90

Northern Electric

**626 'Glasgow', c.1938**
5 tubes, 2 bands
$125

Northern Electric

**722 'Wellington, c.1937**
7 tubes, 3 bands
$225

Northern Electric

**B558, c.1940**
5 tubes, 5 bands, DC
$175

**NORDEN-HAUCK**
**SUPER DX5, c.1931**
5 tubes, allwave
$150

**NPC**
**40A, c.1939**
5 tubes, 1 band
$125

**NPC**
**415, c.1939**
5 tubes, 1 band
$125

**NPC**
**455L, c.1939**
5 tubes, 1 band, radio-phono
$35

**NPC**
**D19, c.1942**
6 tubes, 1 band
$75

**NPC**
**L47-1, c.1942**
5 tubes, 1 band, AC/DC
$225

**NPC**
**L49, c.1942**
5 tubes, 1 band
$225

**NPC**
**L51, c.1942**
5 tubes, 1 band, DC
$25

**NPC**
**L448, c.1942**
5 tubes, 1 band, radio-phono
$125

NPC

**L553C, c.1942**
5 tubes, 1 band, radio-phono
$35

NPC

**LG36, c.1942**
phono remote
$25

NPC

**'Space-Mizer', c.1942**
5 tubes, 1 band, AC/DC
$75

NPC

**420A, c.1939**
5 tubes, 1 band
$35

Olympic

**6-501, c.1946**
5 tubes, 1 band
brown $125, black $150, ivory $150

Olympic

**6-504, c.1946**
5 tubes, 1 band, radio-phono
$75

Olympic

**6-505, c.1946**
5 tubes, 1 band
black $100, ivory $125

Olympic

**6-506, c.1946**
5 tubes, 1 band
$50

Olympic

**6-507, c.1946**
5 tubes, 1 band, radio-phono
$45

Olympic

**6-601, c.1946**
5 tubes, 1 band
black $50, ivory $60

Olympic

**6-602, c.1946**
6 tubes, 1 band
$65

Oriole Golden Voiced Radio

**102, c.1939**
5 tubes, 1 band
$75

Oriole Golden Voiced Radio

**120, c.1939**
6 tubes, 2 bands, radio-phono
$125

Oriole Golden Voiced Radio

**1020, c.1939**
7 tubes, 2 bands, radio-phono
$125

Oriole Golden Voiced Radio

**1224, c.1939**
5 tubes, 2 bands, DC
$125

Oriole Golden Voiced Radio

**1112, c.1939**
7 tubes, 2 bands, DC
$50

Oriole Golden Voiced Radio

**3103, c.1939**
6 tubes, 1 band
$75

**C3-104, c.1939**
6 tubes, 1 band
$75

**R2825,27, c.1942**
5 tubes, 1 band, AC/DC
R2825 - brown $60
R2827 - ivory $75

**R2828,31, c.1942**
5 tubes, 1 band, AC/DC
R2828 - brown $100
R2831 - ivory $125

**R2838/9, c.1942**
6 tubes, 2 bands, AC/DC
R2838 - brown $50
R2839 - ivory $60

**R2868, c.1942**
5 tubes, 1 band, DC
$20

**R3105, c.1942**
5 tubes, 1 band, radio-phono
$35

**R3111, c.1942**
4 tubes, 1 band, DC
$75

**R3123, c.1942**
7 tubes, 2 bands, AC/DC
$65

**R3199/3200, c.1942**
5/6 tubes, 2 bands, radio-phono
$60

Oriole

**R3202, c.1942**
8 tubes, FM Converter
$40

Oriole

**R4140, c.1942**
6 tubes, 4 bands
$50

Oriole

**S134, c.1942**
5 tubes, 1 band, AC/DC
$25

Pacific

**23C3, c.1938**
6 tubes, 3 bands, DC
$75

Packard-Bell

**'Kompak', c.1938**
5 tubes, 1 band
brown $150, black $175, red $750

Paramount

**705G, c.1935**
5 tubes, 2 bands
$150

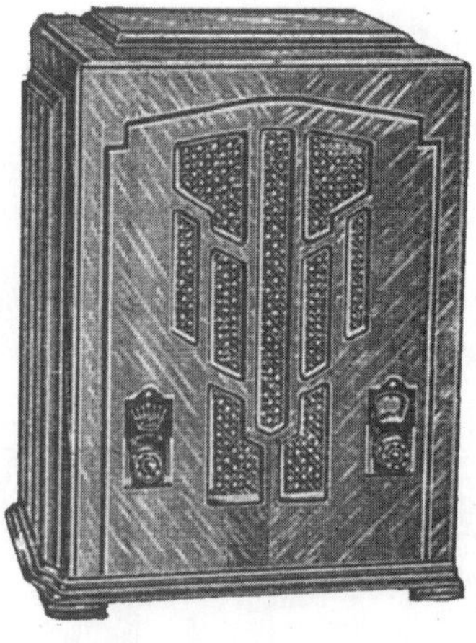

Paramount

**705J, c.1935**
5 tubes, 2 bands
$350

Paramount

**706H, c.1935**
5 tubes, 1 band
$75

Paramount

**900K, c.1935**
5 tubes, 2 bands
$125

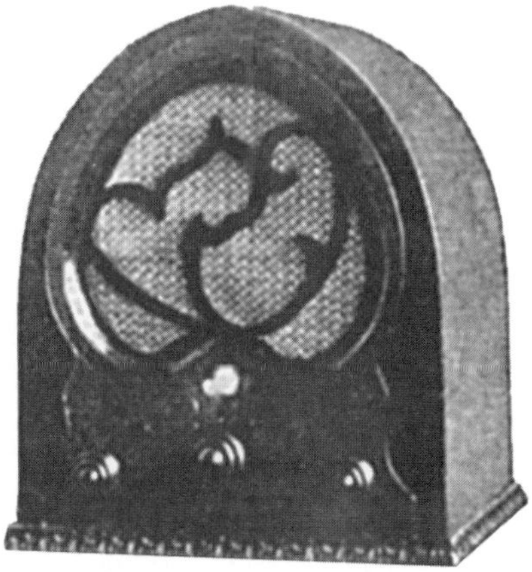

Paramount

**'Kewpie', c.1932**
4 tubes, 1 band
$1200

Paramount

**P-M/P-F, c.1935**
5 tubes, 2 bands
P-M (AC) $225, P-F (DC) $110

Paramount

**PMR, c.1935**
6 tubes, 1 band
$125

Paramount

**V 'UNIVERSAL', c.1934**
4 tubes, 1 band
$110

Paramount

**W, c.1935**
8 tubes, 5 bands
$375

Pathe

**UNIVERSAL, c.1935**
5 tubes, 1 band
$125

PATTERSON

**86AW, c.1935**
8 tubes, 4 bands
$650

PATTERSON

**PR16, c.1934**
16 tubes, 5 bands, bandspread
$750

Perry

**'Deluxe Midget', c.1931**
7 tubes, 1 band
$350

Perry

**Jr. Gothic, c.1931**
5 tubes, 1 band
$400

Perry

**Junior, c.1931**
4 tubes, 1 band
$300

PHILCO CANADA

**57, c,1949**
5 tubes, 1 band
brown $40, ivory $45

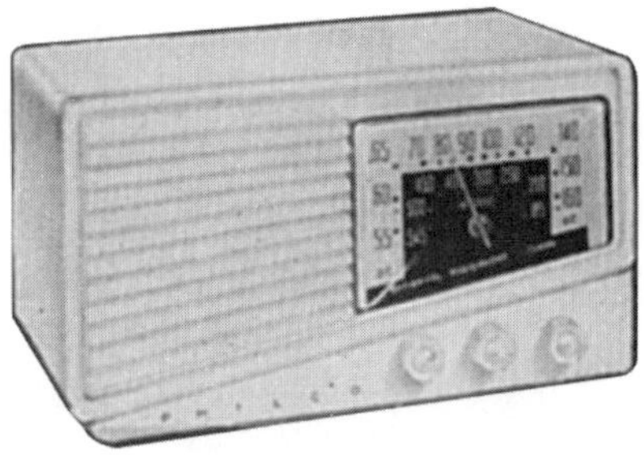

PHILCO CANADA

**82, c.1949**
5 tubes, 1 band
brown $30, ivory $35

PHILCO CANADA

**82, c.1949**
5 tubes, 1 band
brown $45, ivory $50

PHILCO CANADA

**84, c.1949**
5 tubes, 1 band
$30

PHILCO CANADA

**85, c.1949**
5 tubes, 1 band
brown $125, ivory $135

PHILCO CANADA

**86, c.1949**
7 tubes, AM/FM
$20

PHILCO CANADA

**217, c.1949**
4 tubes, 1 band, DC
$25

PHILCO CANADA

**222, c.1949**
5 tubes, 1 band, DC
$15

PHILCO CANADA

**411, c.1949**
4 tubes, 1 band, DC
$30

PHILCO CANADA

**412, c.1949**
5 tubes, 1 band, AC/DC
$35

PHILCO CANADA

**623T/624T 'Bullet', c.1938**
6 tubes, 3 bands
$175

PHILCO CANADA

**C7T, c.1938**
6 tubes, 2 bands
$165

PHILCO CANADA

**C9T, c.1938**
6 tubes, 2 bands
$135

PHILCO CANADA

**C10T,C325T 'Bullet', c.1938**
5 tubes, 2 bands
$175

PHILCO
CANADA

**M15, c.1949**
phono remote
$20

PHILCO
CANADA

**6C/CI, c.1939**
5 tubes, 1 band
6C natural $50, 6CI ivory $60

PHILCO
CANADA

**7C, c.1939**
5 tubes, 1 band
65

PHILCO

**7T, c.1939**
5 tubes, 1 band
$50

PHILCO

**10, 'Bullet', c.1938**
5 tubes, 1 band
$175

PHILCO

**30T, c.1939**
6 tubes, 2 bands
$125

PHILCO

**145T, c.1940**
6 tubes, 2 bands
$110

PHILCO

**224, c.1949**
5 tubes, 1 band, DC
$20

PHILCO

**502, c.1949**
5 tubes, 1 band, radio-phono
$50

PHILCO
**543/545, c.1953**
5 tubes, 1 band
$125

PHILCO
**547, c.1953**
5 tubes, 1 band
$50

PHILCO
**550, c.1953**
5 tubes, 1 band
$30

PHILCO
**560, c.1953**
5 tubes, 1 band
$25

PHILCO
**561, c.1953**
5 tubes, 1 band
$60

PHILCO
**562, c.1953**
5 tubes, 1 band
$35

PHILCO

**563, c.1953**
5 tubes, 1 band
black $125, red $165

PHILCO

**564, c.1953**
5 tubes, 1 band
$50

PHILCO

**566, c.1953**
5 tubes, 1 band
$165

PHILCO

**700/701, c.1953**
5 tubes, 1 band
$40

PHILCO

**702, c.1953**
5 tubes, 1 band
$25

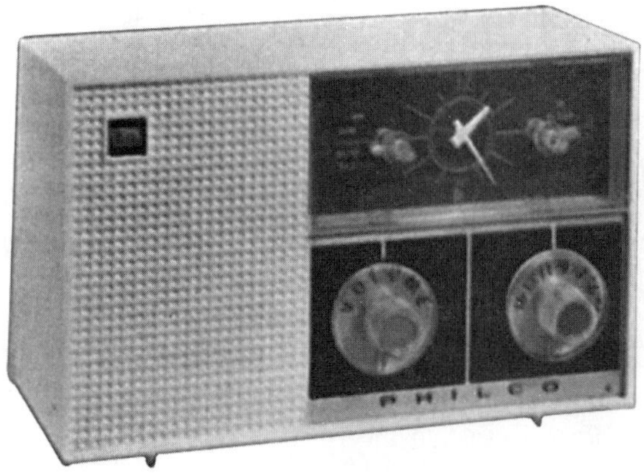


PHILCO

**709, c.1964**
5 tubes, 1 band, clock-radio
black $40, white $45

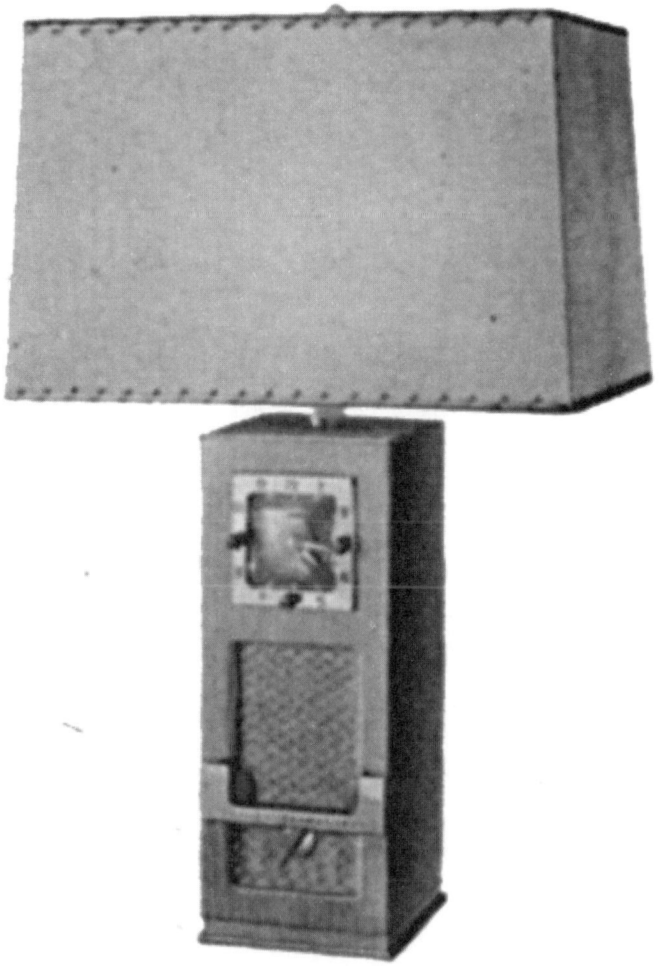

PHILCO

**706, c.1953**
5 tubes, 1 band, clock-radio-lamp
$225 with shade, $150 without shade

PHILCO

**707, c.1953**
5 tubes, 1 band, clock-radio
$250

PHILCO

**656, c.1953**
5 tubes, 1 band
$40

PHILCO

**714, c.1964**
5 tubes, 1 band, clock-radio
tan $45, pink $65

PHILCO

**715, c.1964**
5 tubes, 1 band, clock-radio
tan $45, aqua $65

PHILCO

**712, c.1964**
5 tubes, 1 band, clock-radio
white $35, black $35

PHILCO

**716, c.1964**
5 tubes, 1 band, clock-radio
tan $35, aqua $50

PHILCO

**719, c.1964**
9 tubes, AM/FM, clock-radio
brown $25, aqua $45

PHILCO

**762, c.1964**
5 tubes, 1 band, clock-radio
pink $65, green $65, ivory $35

PHILCO

**763, c.1960**
5 tubes, 1 band, clock-radio
black $35, ivory $35

PHILCO

**764, c.1960**
5 tubes, 1 band, clock-radio
ivory $40, pink $65

PHILCO
**765, c.1960**
5 tubes, 1 band, clock-radio
black $250, ivory $225

PHILCO
**788, c.1964**
9 tubes, AM/FM, clock-radio
$25

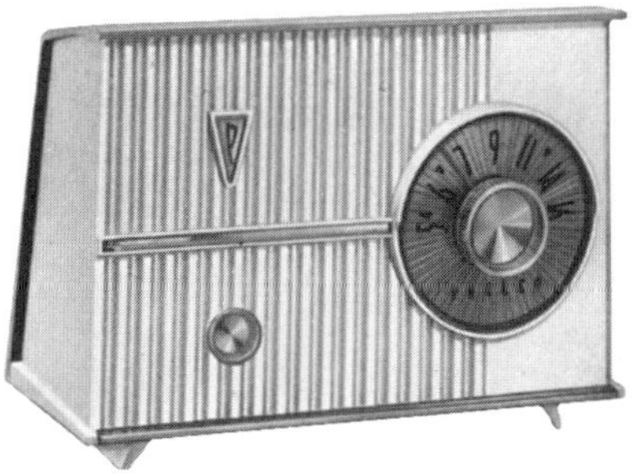

PHILCO
**834, c.1960**
5 tubes, 1 band
pink $75, aqua $85

PHILCO
**835, c.1960**
5 tubes, 1 band
red $125, aqua $85, pink $75

PHILCO
**869, c.1964**
5 tubes, 1 band
ivory $35

PHILCO
**872, c.1964**
5 tubes, 1 band
tan $45, aqua $65

PHILCO
**874, c.1964**
5 tubes, 1 band
brown $20, blue $45

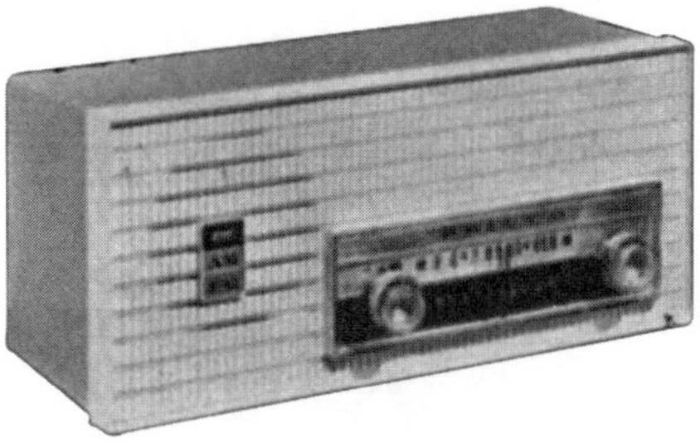

PHILCO
**936, c.1964**
9 tubes, AM/FM
tan $25, white $30

PHILCO
**937, c.1964**
9 tubes, AM/FM
brown $20, aqua $35

PHILCO
**938, c.1964**
9 tubes, AM/FM
$35

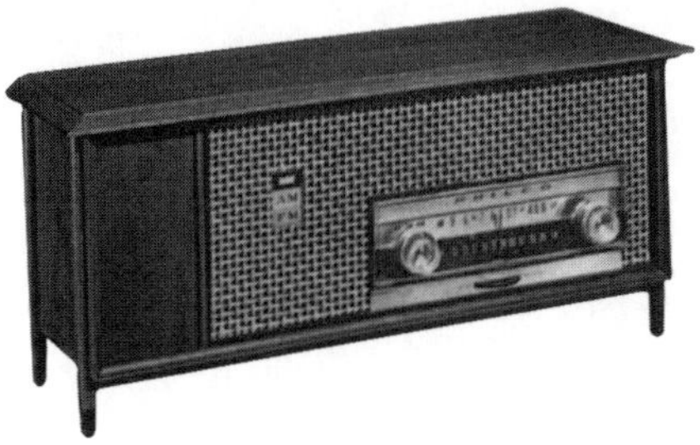

PHILCO
**939, c.1964**
9 tubes, AM/FM
$65

PHILCO
**956, c.1953**
7 tubes, AM/FM
$45

PHILCO
**984, c.1960**
9 tubes, AM/FM
black $25, aqua $45

PHILCO
**1350, c.1953**
5 tubes, 1 band, radio-phono
$45

PHILCO
**C12T/C24T, c.1938**
6 tubes, 2 bands
$125

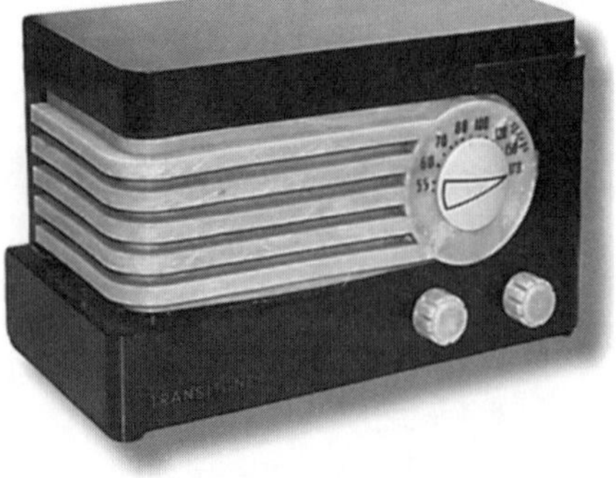

PHILCO
**TP7, c.1939**
5 tubes, 1 band
$350

**PHILMORE**
**1-Tube, c.1932**
1 tube, 1 band
$125

*Phoenix*
**'Miniature, c.1931**
4 tubes, 1 band
$225

Phonola

**'Audette', c.1941**
phono remote
$35

Phonola

**'Collegiate', c.1941**
5 tubes, 1 band
$65

Phonola

**'Comet', c.1941**
5 tubes, 2 bands, DC
$35

Phonola

**'Express', c.1941**
4 tubes, 1 band, DC
$35

Phonola

**'Lucerne', 1941**
5 tubes, 1 band, DC
$35

Phonola

**'Falcon', c.1941**
5 tubes, 1 band
$60

Phonola

**'Melody', c.1941**
5 tubes, 1 band, radio-phono
$60

Phonola

**PBH51, c.1938**
5 tubes, 2 bands, DC
$75

Phonola

**PBK72/PH51/PBH51, c.1938**
PBK72 - 7 tubes, 2 bands, DC $60
PH51 - 5 tubes, 1 band, AC $110
PBH51 - 5 tubes, 1 band, DC $45

Phonola

**PK52/72, c.1938**
PK52 - 5 tubes, 2 bands, AC $100
PK72 - 5 tubes, 2 bands, DC $40

Phonola

**PK52/PK72/PV162, c.1938**
PK52 - 5 tubes, 2 bands, AC $150
PK72 - 5 tubes, 2 bands, DC $75
PV162 - 6 tubes, 2 bands, AC $150

Phonola

**'Post', c.1941**
4 tubes, 1 band, DC
$50

Phonola

**'Prom', c.1941**
phono remote
$30

Phonola

**'Rambler', c.1941**
PK72 - 5 tubes, 1 band
$60

Phonola

**'Stafford, c.1941/6/5**
6 tubes, 5 bands, DC
$35

Phonola

**'Topic', c.1941**
5 tubes, 1 band
$65

Phonola

**'Troubador', c.1941**
5 tubes, 1 band
$90

Phonola

**'Windsor', c.1941**
5 tubes, 2 bands, DC
$35

Pierce-Aero

**724, c.1930**
5 tubes, 1 band
$60

Pierce-Aero

**'DeWald' 1, c.1933**
5 tubes, 1 band
$90

PILOT Radio

**BG562, c.1938**
7 tubes, 2 bands
$1500

PILOT Radio

**'Duo', c.1933**
4 tubes, 1 band
$250

PILOT Radio

**E20, c.1933**
5 tubes, 1 band
$275

PILOT Radio

**'Broadcast', c.1933**
6 tubes, 1 band
$300

PILOT Radio

**'Midget', c.1933**
6 tubes, 1 band
$375

PILOT Radio

**'Midget', c.1931**
7 tubes, 1 band
$400

PILOT Radio

**PE6, c.1930**
6 tubes, 1 band
$125

PILOT Radio

**'Table', c.1931**
6 tubes, 1 band
$250

PILOT Radio

**T91, c.1940**
6 tubes, 2 bands
$110

PILOT Radio

**c.1938**
radio-phono
$150

Porto-Matic

**'Consolette', c.1947**
5 tubes, 1 band, radio-phono
$35

**POWERTONE**

**PRS-5 'Senator', c.1932**
5 tubes, 1 band
$225

**PREMIER**

**'Four', c.1933**
4 tubes, 1 band
$175

PRESTO

**D, c.1937**
portable disc recorder
$125

RADI-O-LITE

**'Radi-O-Lite', c.1939**
4 tubes, 1 band
$750

**RADIETTE**

**100, c.1935**
4 tubes, 1 band
$125

**RADIETTE**

**124, c.1935**
4 tubes, 1 band
$90

**RADIETTE**

**154, c.1935**
5 tubes, 2 bands
$75

**RADIETTE**

**156, c.1935**
5 tubes, 2 bands
$125

**RADIETTE**

**'Carriette', c.1932**
pressbaord, 5 tubes, 1 band
$325

**Radio Keg**

**'Radio Keg', c.1933**
5 tubes, 1 band
$350

**RADIO-SIGHT**

**95, c.1940**
5 tubes, 1 band, radio-lamp
$750

**RADIO-SIGHT**

**100, c.1940**
5 tubes, 1 band, radio-lamp
$750

**RADIO-SIGHT**

**135, c.1940**
5 tubes, 1 band, radio-lamp
$750

**RADIO-SIGHT**

**2905, c.1940**
5 tubes, 1 band, radio-lamp
$750

**RADIO-SIGHT**

**2915, c.1940**
5 tubes, 1 band, radio-lamp
$750

**RADIO-SIGHT**

**3000, c.1940**
5 tubes, 1 band, radio-lamp
$750

**RADIO-SIGHT**

**395, c.1940**
5 tubes, 1 band, radio-lamp
$750

**RADIO-SIGHT**

**500, c.1940**
5 tubes, 1 band, radio-lamp
$750

**RADIETTE**

**'Mantle 9', c.1931**
9 tubes, 1 band
$350

RADOLEK

**10950 'Universal', c.1934**
4 tubes, 1 band
$200

RADOLEK

**10953 'Companion', c.1934**
3 tubes, 1 band
$175

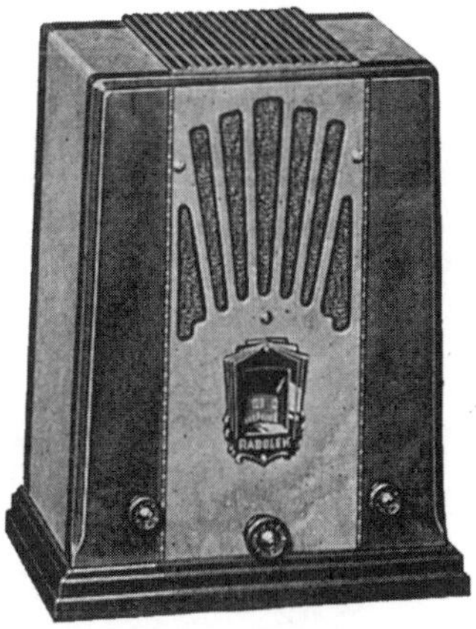

RADOLEK

**10962, c.1934**
4 tubes, 2 bands
$400

RADOLEK

**10963, c.1934**
5 tubes, 2 bands
$225

RADOLEK

**10966, c.1934**
5 tubes, 1 band
$300

RADOLEK

**10967, c.1934**
5 tubes, 2 bands
$225

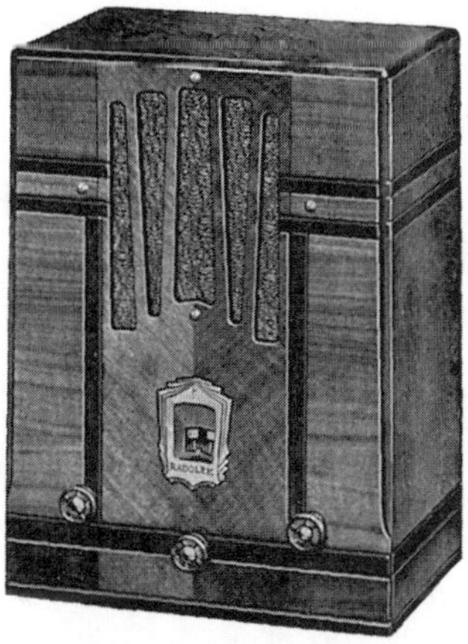

RADOLEK

**10968/9, c.1934**
5 tubes, 2 bands, DC
$100

RADOLEK

**10970, c.1934**
8 tubes, 5 bands
$450

RADOLEK

**10980, c.1934**
6 tubes, 3 bands
$425

RADOLEK

**16742, c.1937**
6 tubes, 3 bands, DC
$125

RADOLEK

**17511, c.1938**
5 tubes, 1 band, AC/DC
$50

RADOLEK

**17510, c.1938**
6 tubes, 2 bands
$175

RADOLEK

**17515, c.1938**
7 tubes, 3 bands
$150

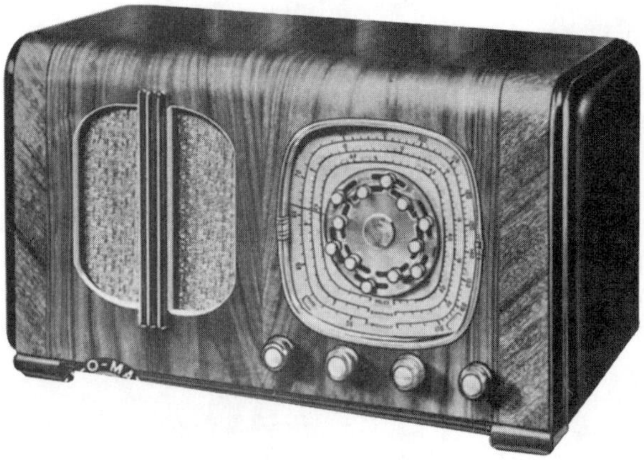

RADOLEK

**17516, c.1938**
8 tubes, 3 bands
$175

RADOLEK

**17518, c.1938**
5 tubes, 1 band
$125

RADOLEK

**17512, c.1938**
6 tubes, 1 band, AC/DC
$50

RADOLEK

**17520, c.1938**
5 tubes, 1 band
$90

RADOLEK

**17521, c.1938**
6 tubes, 2 bands, AC/DC
$125

RADOLEK

**17522, c.1938**
6 tubes, 2 bands
$90

RADOLEK

**17524, c.1938**
5 tubes, 2 bands, AC/DC
$100

RADOLEK

**17526, c.1938**
6 tubes, 2 bands
$125

RADOLEK

**17526, c.1938**
6 tubes, 2 bands
$1200

RADOLEK

**17527, c.1938**
7 tubes, 2 bands, AC/DC
$325

RADOLEK

**17530, c.1938**
7 tubes, 3 bands, DC
$60

RADOLEK

**17532, c.1938**
5 tubes, 1 band, DC
$75

RADOLEK

**17533, c.1938**
5 tubes, 2 bands, DC
$50

RADOLEK

**17540, c.1938**
5 tubes, 2 bands, AC/DC
$75

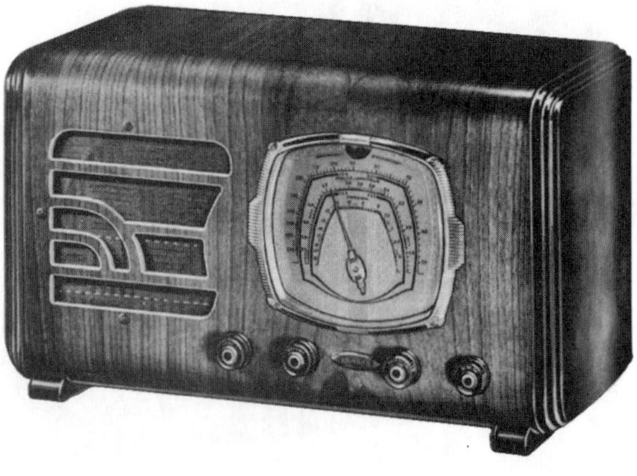

RADOLEK

**17541, c.1938**
7 tubes, 3 bands
$75

RADOLEK

**17567, c.1938**
5 tubes, 1 band, AC/DC
$110

RADOLEK
**17565, c.1938**
11 tubes, 3 bands, AC/DC
$375

RADOLEK
**17568, c.1938**
6 tubes, 2 bands, AC/DC
$65

RADOLEK
**17651, c.1942**
5 tubes, 1 band, radio-phono
$50

RADOLEK
**17664/5, c.1942**
phono remote
$30

RADOLEK
**17666/7, c.1942**
phono remote
$30

RADOLEK
**17668/9, c.1942**
phono remote
$30

RADOLEK
**17699, c.1942**
FM converter
$40

RADOLEK
**C149, c.1940**
5 tubes, 1 band, DC
$40

RADOLEK
**C17517, c.1940**
5 tubes, 1 band
$75

RADOLEK

**C17515, c.1940**
5 tubes, 1 band
$425

RADOLEK

**C17516, c.1940**
6 tubes, 1 band
$150

RADOLEK

**C17518, c.1940**
5 tubes, 1 band, radio-phono
$40

RADOLEK

**C17520/1, c.1940**
5 tubes, 1 band
C17520 - brown $110
C17521 - ivory $125

RADOLEK

**C17526, c.1940**
5 tubes, 1 band
$75

RADOLEK

**C17573/4, c.1940**
6 tubes, 1 band
C17673 - brown $150
C17674 - ivory $175

RADOLEK

**C17657/8, c.1940**
5 tubes, 1 band
C17657 - brown $350
C17658 - ivory $400

RADOLEK

**'Deluxe', c.1942**
crystal
$175

RADOLEK

**'Lowloss', c.1942**
crystal
$100

CANADA

**86T, c.1938**
6 tubes, 3 bands
$75

CANADA

**88T, c.1938**
8 tubes, 3 bands
$125

CANADA

**94BT1, c.1939**
4 tubes, 1 band, DC
$65

CANADA

**96BT, c.1937**
5 tubes, 1 band, DC
$45

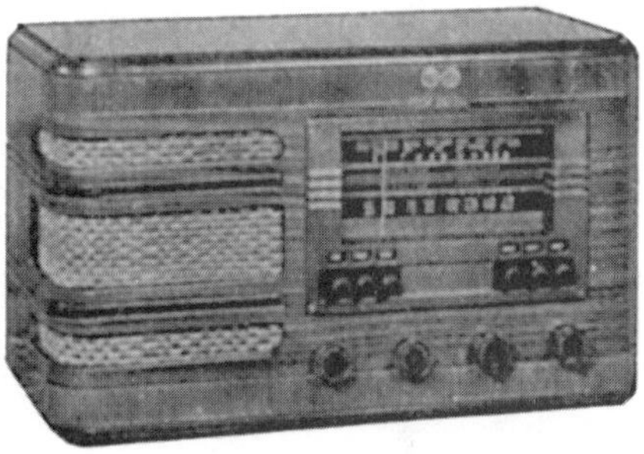

CANADA

**A1, c.1939**
5 tubes, 2 bands
$75

CANADA

**A3, c.1939**
7 tubes, 2 bands
$80

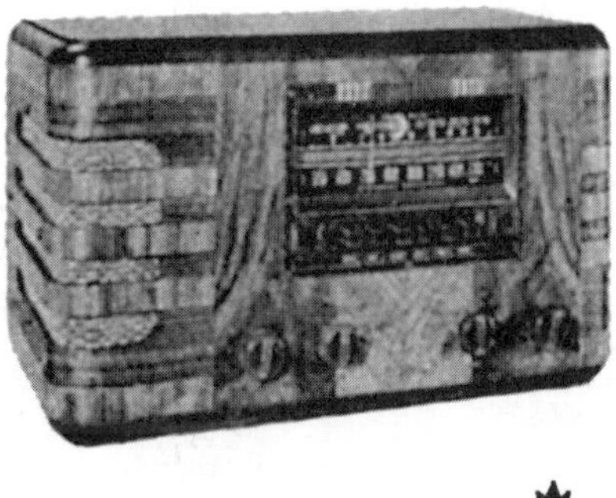

CANADA

**A5, c.1939**
6 tubes, 2 bands
$75

CANADA

**A20, c.1940**
5 tubes, 3 bands
$350

CANADA

**A21, c.1940**
5 tubes, 2 bands
$50

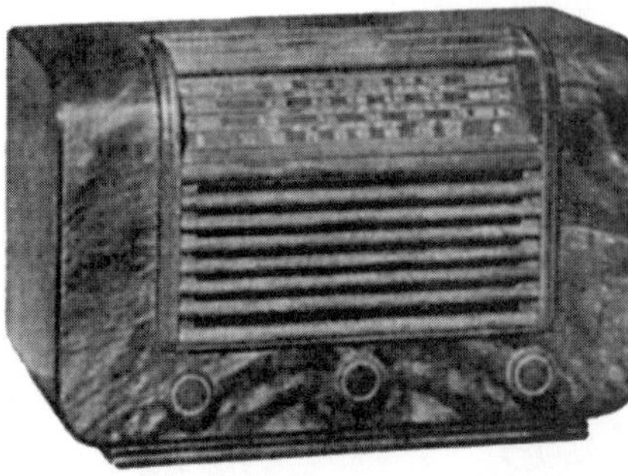

CANADA

**A23, c.1941**
5 tubes, 4 bands
$40

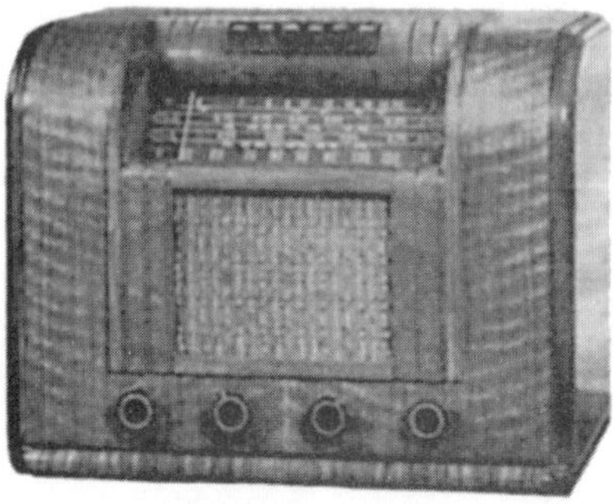

CANADA

**A24, c.1941**
6 tubes, 4 bands
$50

CANADA

**A22, c.1940**
7 tubes, 6 bands
$275

CANADA

**B5/6B5, c.1939**
7 tubes, 5 bands, DC
$30

CANADA

**A25, c.1941**
7 tubes, 6 bands
$275

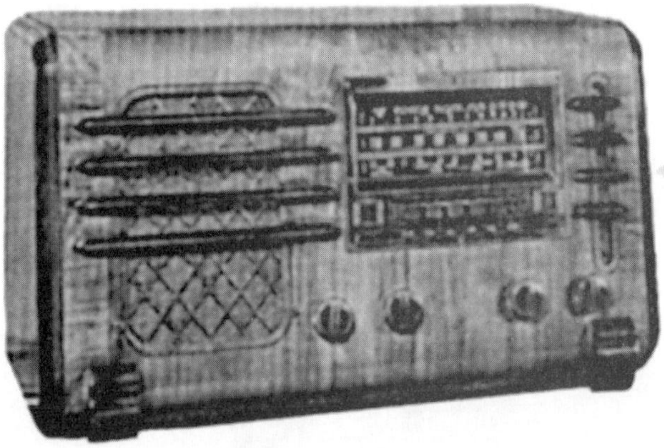

CANADA

**B3/6B3, c.1939**
6 tubes, 3 bands, DC
$65

RCA CANADA

**B1, c.1939**
6 tubes, 1 band, DC
$125

RCA CANADA

**'Ace', c.1939**
5 tubes, 1 band
$75

RCA CANADA

**B60, c.1940**
6 tubes, 1 band, DC
$25

RCA CANADA

**B61, c.1940**
6 tubes, 5 bands, DC
$20

RCA CANADA

**B62, c.1939**
4 tubes, 1 band, DC
$45

RCA CANADA

**B63, c.1941**
4 tubes, 3 bands, DC
$35

RCA CANADA

**B64, c.1941**
4 tubes, 1 band, DC
$50

RCA CANADA

**B67, c.1941**
6 tubes, 5 bands, DC
$25

RCA CANADA

**'Little Nipper', c.1940**
6 tubes, 1 band
$90

CANADA

**'Mascot', c.1941**
5 tubes, 1 band
$75

CANADA

**'Master Nipper', c.1940**
5 tubes, 1 band
$65

CANADA

**'Nipper A', c.1941**
5 tubes, 1 band
$50

CANADA

**'Nipper', c.1939**
5 tubes, 1 band
$110

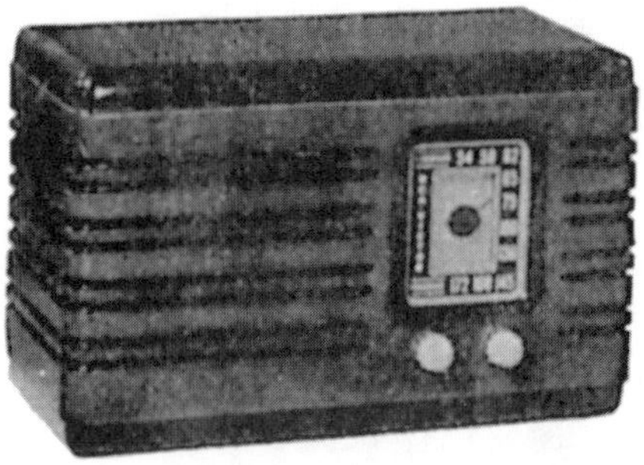

CANADA

**'Nipperette A', c.1939**
4 tubes, 1 band, DC
$60

CANADA

**'Nipperette', c.1939**
4 tubes, 1 band, DC
$75

CANADA

**Q22, c.1939**
6 tubes, 2 bands
$40

CANADA

**R96, c.1940**
3 tubes, phono only
$30

CANADA

**R96A, c.1941**
3 tubes, phono only
$30

RCA CANADA

**R97, c.1941**
3 tubes, phono only
$50

RCA CANADA

**U60A, c.1939**
6 tubes, 3 bands, radio-phono
$60

RCA CANADA

**V1, c.1939**
phono remote
$25

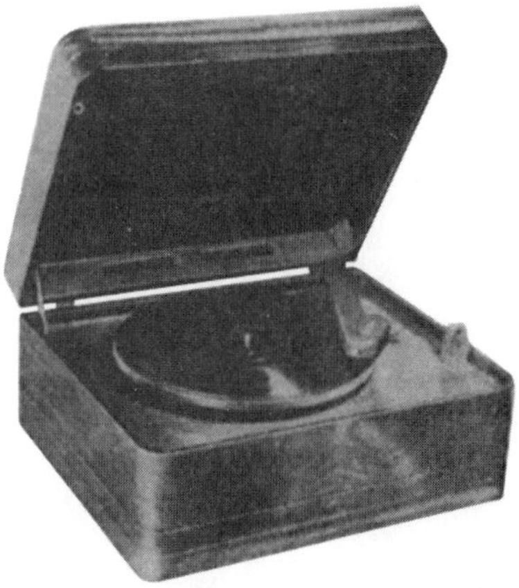

RCA CANADA

**V2, c.1939**
phono remote
$30

RCA CANADA

**V3/V3A, c.1941**
manual wind phono remote
$50

RCA CANADA

**V4, c.1941**
phono remote
$30

RCA CANADA

**V5, c.1941**
phono remote
$30

RCA CANADA

**VA15, c.1941**
phono remote
$40

RCA CANADA

**'Victorette', c.1939**
5 tubes, 1 band
$35

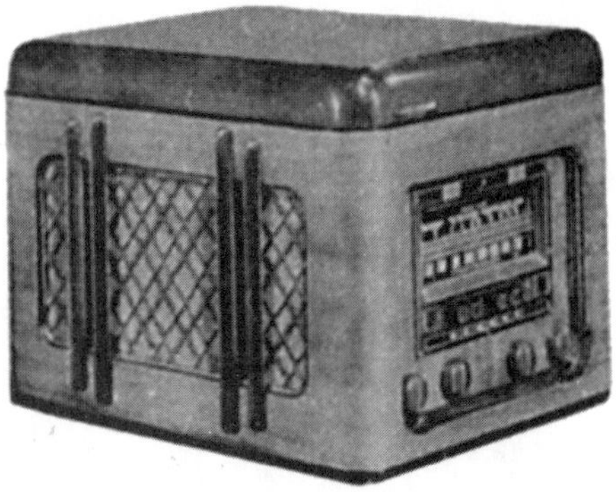

RCA CANADA

**VR1, c.1939**
5 tubes, 2 bands, radio-phono
$65

RCA CANADA

**VR3, c.1939**
5 tubes, 1 band, radio-phono
$35

RCA CANADA

**VR40, c.1940**
5 tubes, 1 band, radio-phono
$50

RCA CANADA

**VR41, c.1940**
5 tubes, 1 band, radio-phono
$50

RCA CANADA

**VR42, c.1940**
5 tubes, 2 bands, radio-phono
$50

RCA CANADA

**VR44, c.1941**
5 tubes, 4 bands, radio-phono
$60

RCA CANADA

**VR44C, c.1941**
5 tubes, 4 bands, radio-phono
$50

RCA Victor

**2BX63 'Globetrotter', c.1953**
5 tubes, 1 band
$25

RCA Victor

**2C511-14, c.1953**
5 tubes, 1 band
2C511 - black $50, 2C512 - ivory $40,
2C513 - red $125, 2C514 - tan $45

RCA Victor
**2ES3, c.1952**
3 tubes, phono only
$50

RCA Victor
**2G511, c.1953**
5 tubes, 1 band
black $50, ivory $40,
red $125, grey $50

RCA Victor
**2R51, c.1953**
5 tubes, 1 band
black & tan $125
tan & ivory $110

RCA Victor
**2US7, c.1952**
5 tubes, 1 band, radio-phono
$35

RCA Victor
**2XF91, c.1953**
5 tubes, 1 band, clock-radio
maroon $50, ivory $35, green $50
red $90, tan $40

RCA Victor
**4QB3/4QB3X, c.1949**
4 tubes, 3 bands, DC
$20

RCA Victor
**7Q51X, c.1949**
7 tubes, 3 bands, bandspread
$60

RCA Victor
**6PQ3, c.1949**
5 tubes, 3 bands, AC/DC
$75

RCA Victor
**8X541/2/4/7, c.1949**
5 tubes, 1 band
$60

RCA Victor
**7QB, c.1939**
7 tubes, 3 bands
$75

RCA Victor
**9BX5, c.1949**
5 tubes, 1 band, AC/DC
$45

RCA Victor
**9EY3/M3, c.1949**
3 tubes, phono only
$50

RCA Victor
**9EY35 'Disney', c.1949**
3 tubes, phono only
$250

RCA Victor
**9JY/JYM, c.1949**
3 tubes, phono remote
$40

RCA Victor
**9EY36 'Roy Rogers'**
3 tubes, phono only
$250

RCA Victor
**9Q53, c.1949**
8 tubes, 3 bands, bandspread
$60

RCA Victor
**9X561/2, c.1949**
5 tubes, 1 band
9X561 - brown $75
9X562 - ivory $85

RCA Victor

**9X641/2, c.1949**
6 tubes, 1 band
9X641 - brown $35
9X642 - ivory $40

RCA Victor

**9X651/2, c.1949**
6 tubes, 1 band
9X651 - brown $35
9X652 - ivory $40

RCA Victor

**9Y51/510/511, c.1949**
5 tubes, 1 band, radio-phono
$50

RCA Victor

**28B 'Chest', c.1933**
5 tubes, 1 band
$150

RCA Victor

**45EY2/3, c.1950**
3 tubes, phono only
$60

RCA Victor

**45EY3, c.1950**
3 tubes, phono only
$75

RCA Victor

**28D 'Tambour', c.1933**
5 tubes, 1 band
$250

RCA Victor

**28E 'Sheraton', c.1933**
5 tubes, 1 band
$150

*RCA Victor*

**55F, c.1942**
5 tubes, 1 band, DC
$40

*RCA Victor*

**55U/AU, c.1942**
5 tubes, 1 band, radio-phono
$50

*RCA Victor*

**54B1/2/3, c.1942**
4 tubes, 1 band, DC
$45

*RCA Victor*

**45EY4, c.1952**
3 tubes, 1 band, phono only
$40

*RCA Victor*

**56X, c.1942**
6 tubes, 1 band
$25

*RCA Victor*

**56X10, c.1942**
6 tubes, 2 bands
$25

*RCA Victor*

**56X11, c.1942**
6 tubes, 2 bands
$25

*RCA Victor*

**56X2, c.1942**
6 tubes, 1 band
$25

RCA Victor

**56X3, c.1942**
6 tubes, 1 band
$50

RCA Victor

**56X5, c.1942**
6 tubes, 2 bands
$50

RCA Victor

**616/7, c.1942**
5 tubes, 2 bands
616 - brown $25, 617 - ivory $30

RCA Victor

**63E/M, c.1942**
3 tubes, phono only
$40

RCA Victor

**64F1/2, c.1942**
5 tubes, 1 band, DC
$20

RCA Victor

**64F3, c.1942**
5 tubes, 1 band, DC
$20

RCA Victor

**65F, c.1942**
5 tubes, 1 band, DC
$20

RCA Victor

**65U/AU, c.1942**
5 tubes, 1 band, radio-phono, DC
$20

RCA Victor

**65U-1, c.1942**
5 tubes, 1 band, radio-phono, DC
$20

**RCA Victor**

**65X1/2, c.1942**
5 tubes, 1 band
65X1 - brown $30
65X2 - ivory $35

**RCA Victor**

**65X8/9, c.1942**
5 tubes, 1 band
$25

**RCA Victor**

**66X13, c.1942**
6 tubes, 1 band
$30

**RCA Victor**

**68R3, c.1942**
8 tubes, 2 bands
$30

**RCA Victor**

**68R4, c.1942**
8 tubes, 2 bands
$40

**RCA Victor**

**96T5, c.1939**
6 tubes, 2 bands
$115

**RCA Victor**

**96T6, c.1939**
6 tubes, 2 bands
$35

**RCA Victor**

**B411, c.1950**
4 tubes, 1 band, DC
marbeled plastic $90

**RCA Victor**

**BT6-10, c.1935**
6 tubes, 2 bands, DC
$75

RCA Victor
**BT7-8, c.1935**
7 tubes, 2 bands, DC
$70

RCA Victor
**BX55, c.1950**
4 tubes, 1 band, AC/DC
$45

RCA Victor
**BX57, c.1950**
4 tubes, 1 band, AC/DC
$45

RCA Victor
**BX6, c.1950**
5 tubes, 1 band, AC/DC
$65

RCA Victor
**Q10/12, c.1942**
5 tubes, 1 band
Q10 - brown $25
Q11 - ivory $30

RCA Victor
**Q22A, c.1942**
6 tubes, 5 bands
$50

RCA Victor
**Q32, c.1942**
6 tubes, 5 bands
$40

RCA Victor
**Q34, c.1942**
8 tubes, 5 bands
$45

RCA Victor
**Q36, c.1942**
11 tube, 7 bands
$75

RCA Victor
**Q103, c.1942**
5 tubes, 3 bands
$35

RCA Victor
**Q121, c.1942**
6 tubes, 5 bands
$35

RCA Victor
**QB55, c.1942**
4 tubes, 3 bands, DC
$15

RCA Victor
**QU72/72A, c.1942**
5 tubes, 2 bands, radio-phono
$35

RCA Victor
**R4, c.1932**
7 tubes, 1 band
$250

RCA Victor
**R8, c.1932**
8 tubes, 1 band
$325

RCA Victor
**R28A, c.1933**
5 tubes, 1 band
$110

RCA Victor
**R70N 'Nursery', c.1933**
7 tubes, 1 band
$900

RCA Victor
**Shortwave Converter, c.1932**
3 tubes
$125

RCA Victor
**R17, c.1933**
4 tubes, 1 band, AC/DC
$225

RCA Victor
**R17M, c.1933**
4 tubes, 1 band, AC/DC
$125

RCA Victor
**R27, c.1933**
5 tubes, 1 band
$175

RCA Victor
**R28C 'Colonial', c.1933**
5 tubes, 1 band
$200

RCA Victor
**R28E, c.1933**
5 tubes, 1 band
$150

RCA Victor
**R28P, c.1933**
5 tubes, 1 band
$250

**RCA Victor**
**R28F/RE40, c.1933**
5 tubes, 1 band, radio-phono
$225

**RCA Victor**
**R225, c.1934**
5 tubes, 1 band, AC/DC
repwood $275

**RCA Victor**
**X551/2, c.1950**
5 tubes, 1 band
X551 - brown $50
X552 - ivory $60

**RCA Victor**
**T8-14/16, c.1935**
8 tubes, 3 bands
$325

**RCA Victor**
**T9-9, c.1935**
9 tubes, 3 bands
$350

**RCA Victor**
**T11-8, c.1935**
8 tubes, 3 bands
$300

**RCA Victor**
**X711, c.1950**
7 tubes, AM/FM
$35

**RCI**
**'Super 6', c.1933**
6 tubes, 1 band
$250

**Recoton**
**'Phono Remote, c.1938**
$50

**RCA Victor**
**X55, c.1939**
5 tubes, 1 band
$75

**RCA Victor**
**X60, c.1939**
5 tubes, 1 band
$65

**Regal**
**P84, c.1941**
phono remote
$25

**Regal**
**R40 'Multi-Mu Midget', c.1932**
4 tubes, 1 band
$250

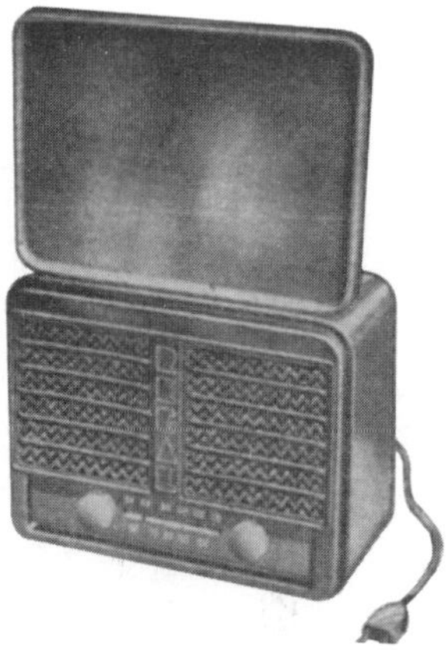

**Regal**
**747 'Super Mite', c.1947**
4 tubes, 1 band
$60

**Regal**
**801, c.1947**
6 tubes, 1 band, AC/DC
$35

**Remler**
**26 'Scottie', c.1934**
5 tubes, 1 band
$750

**Remler**
**46/47, c.1937**
5 tubes, 2 bands
ivory $600, black & ivory $750
beetle $1000

*Remler*

**54, c.1938**
5 tubes, 1 band
brown $250, ivory $350

*Remler*

**61, c.1938**
6 tubes, 1 band
$175

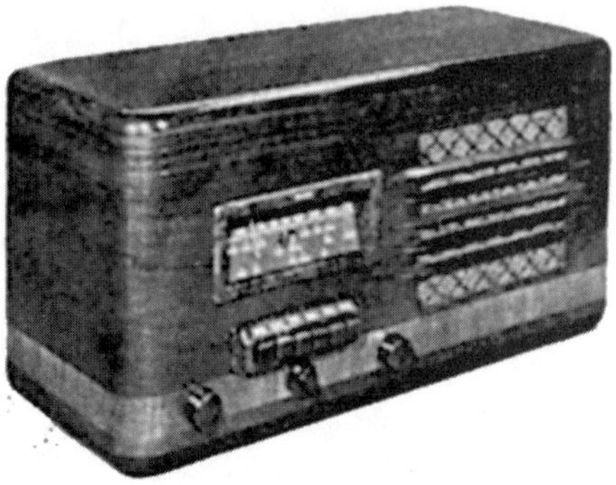

*Remler*

**464, c.1941**
6 tubes, 2 bands
$125

*Remler*

**471 'Micordio', c.1940**
6 tubes, 2 bands, radio-phono-recorder
$150

*Remler*

**'Superheterodyne', c.1931**
8 tubes, 1 band
$400

**RETS**

**1946**
5 tubes, 1 band
brown $50, ivory $75

*Revere*

**'Patrician', c.1932**
5 tubes, 1 band
$2500

**REXON-THORENS**

**Pocket Portable Phono**
manual wind
$500

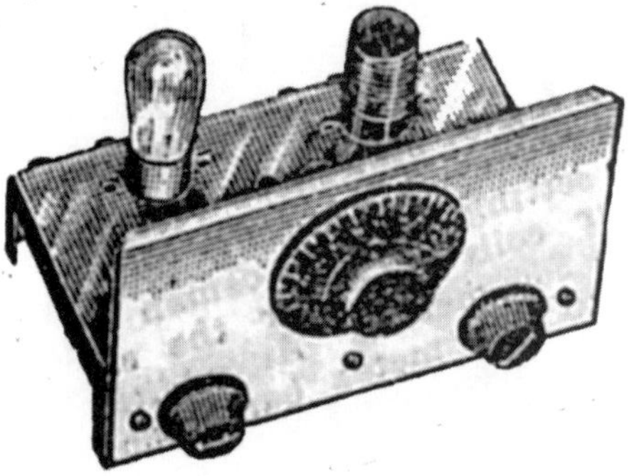

**ROCKET**

**SW5014 'Pal', c.1932**
1 tube, shortwave, DC
$75

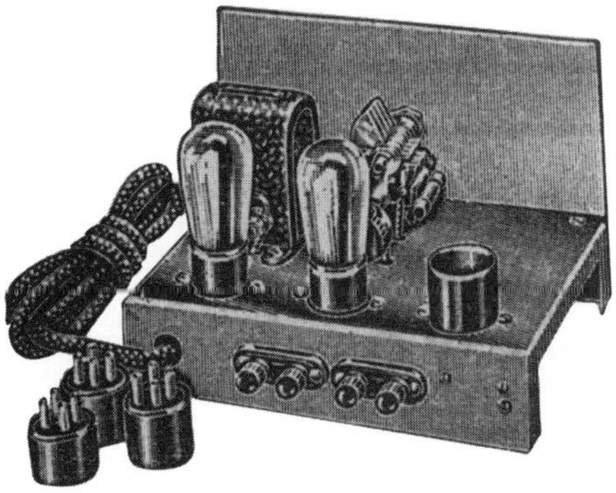

**ROCKET**

**SW5017, c.1935**
2 tubes, shortwave, DC
$75

**ROGERS**

**1051, c.1936**
5 tubes, 2 bands
$500

**ROGERS**

**1140, c.1937**
5 tubes, 1 band
$175

**ROGERS**

**933, c.1937**
9 tubes, 3 bands
$400

**ROGERS**

**1261, c.1938**
6 tubes, 3 bands
$350

**ROGERS**

**1270/7/3, c.1938**
7 tubes, 3 bands
$300

**ROGERS**

**'Scout', c.1937**
5 tubes, 1 band
$275

**Scott**

**FM TUNER, c.1940**
FM converter
$300

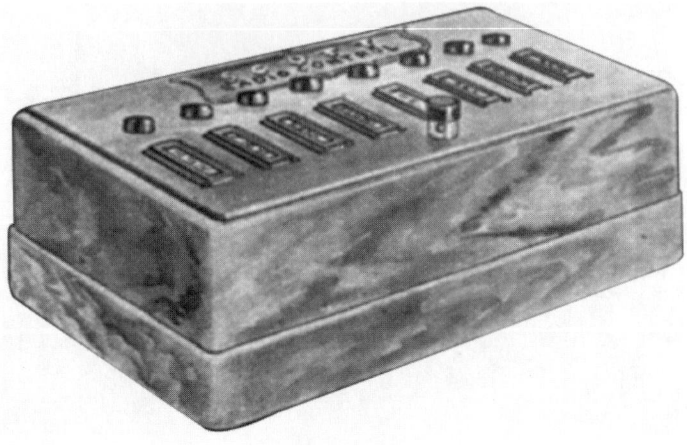

**Scott**

**Remote Control, c.1931**
$250

ROYAL

**'Midget', c.1933**
5 tubes, 1 band
$125

SEE-ALL

**'Televisor Kit', c.1932**
mechanical television converter
$1500

**Sentinel**

**110, c.1932**
7 tubes, 1 band
$150

**Sentinel**

**111, c.1931**
5 tubes, 1 band
$250

**Sentinel**

**116, c.1932**
5 tubes, 1 band
$225

**Sentinel**

**124AT, c.1939**
5 tubes, 1 band
$125

**Sentinel**
**127BT, c.1939**
5 tubes, 1 band, DC
$35

**Sentinel**
**175BTW, c.1940**
5 tubes, 1 band, DC
$20

**Sentinel**
**194ULTO, c.1940**
5 tubes, 1 band
brown $75, ivory $100

**Sentinel**
**204AT/19CB, c.1940**
5 tubes, 1 band, radio-phono
$60

**Sentinel**
**31B74, c.1936**
4 tubes, 1 band
$75

**Sentinel**
**23/63, c.1936**
4/5 tubes, 1 band, DC
$50

**Sentinel**
**571, c.1933**
5 tubes, 1 band, AC/DC
brown $125, green $150

**Sentinel**
**574, c.1933**
5 tubes, 1 band, AC/DC
$175

**Sentinel**
**'Duola', c.1933**
5 tubes, 1 band
$175

**Sentinel**
**'Midget', c.1933**
5 tubes, 1 band
$75

SETCHELL CARLSON
**29/331, c.1939**
5/6 tubes, 1 band, DC
$25

SETCHELL CARLSON
**221/333/621, c.1939**
221 - 7 tubes, 3 bands $65
333 - 9 tubes, 3 bands, DC $50
621 - 5 tubes, 2 bands $55

**SHURA-TONE**
**'Howdy Doody', c.1957**
manual phonograph
$250

**Silver Marshall**
**739, c.1932**
3 tubes, shortwave
$125

**Silver Marshall**
**F4, c.1932**
7 tubes, 1 band
$300

SILVER
**7C7T, c.1940**
7 tubes, 2 bands
$75

SILVER
**9A4, c.1940**
4 tubes, 1 band
$225

SILVER

**100 'Air Cruiser', c.1937**
5 tubes, 1 band
$1000

SILVER

**1310, c.1940**
wireless phono remote
$30

SILVER

**150, c.1937**
4 tubes, 1 band
$110

SILVER

**139, c.1937**
5 tubes, 2 bands
$450

SILVER

**146, c.1937**
6 tubes, 2 bands
$350

SILVER

**147, c.1937**
7 tubes, 3 bands
$125

SILVER

**149, c.1937**
6 tubes, 2 bands
$450

## SILVER

**152, c.1937**
5 tubes, 2 bands
$175

## SILVER

**250, c.1937**
5 tubes, 1 band
$150

## SILVER

**450 'Dwarf', c.1939**
4 tubes, 1 band
$150

## SILVER

**480I, c.1937**
4 tubes, 2 bands, DC
$40

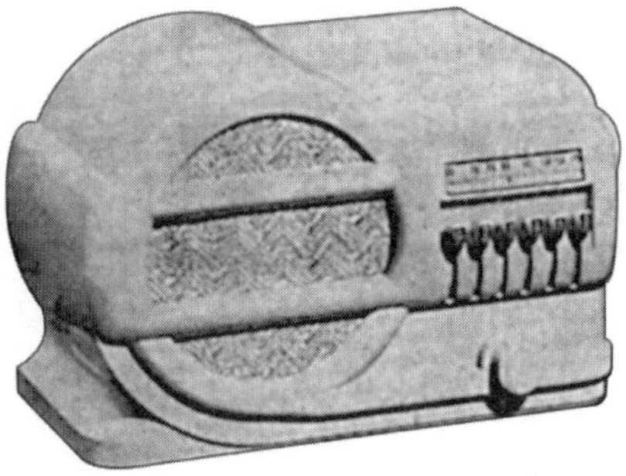

## SILVER

**519, c.1940**
5 tubes, 1 band
brown $350, ivory $400

## SILVER

**553, c.1940**
5 tubes, 1 band
$75

## SILVER

**6141E, c.1937**
7 tubes, 2 bands
$275

## SILVER

**6322XE, c.1937**
8 tubes, 3 bands
$750

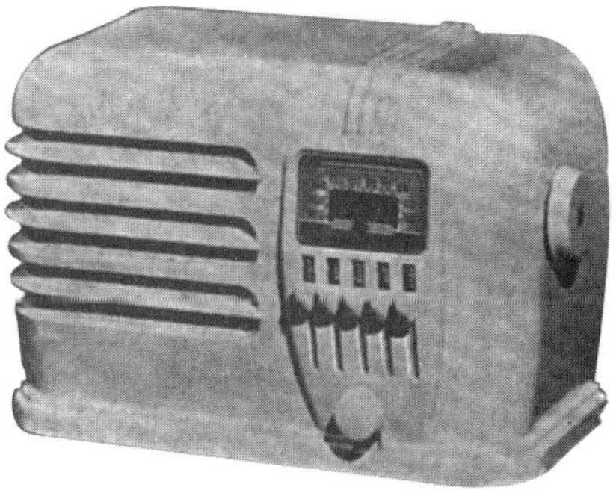

SILVER
**635, c.1940**
6 tubes, 1 band
brown $150, ivory $175

SILVER
**636V, c.1937**
6 tubes, 3 bands, DC
$100

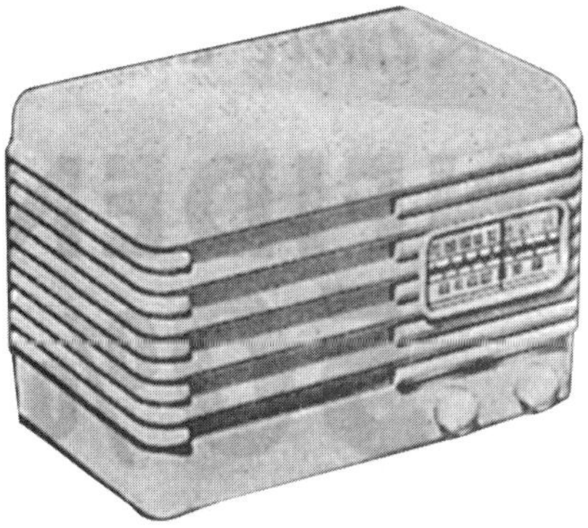

SILVER
**A11, c.1940**
5 tubes, 1 band
$60

SILVER
**B666, c.1940**
6 tubes, 1 band
brown $225, ivory $300

SILVER
**B667P, c.1940**
6 tubes, 1 band
brown $300, ivory $400

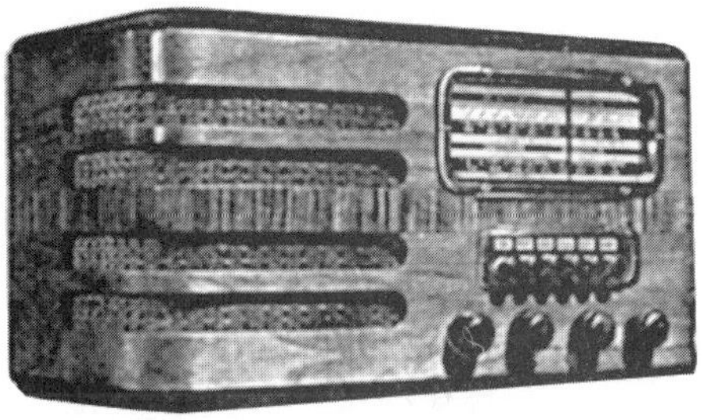

SILVER
**EA33, c.1940**
6 tubes, 2 bands
$75

SILVER
**FA55, c.1940**
7 tubes, 2 bands
$75

SILVER
**M22, c.1940**
4 tubes, 1 band, DC
$50

SILVER
**M8, c.1937**
8 tubes, 3 bands
$375

SILVER

**FA55, c.1940**
7 tubes, 2 bands
$225

SILVER

**ME17, c.1937**
8 tubes, 3 bands
$225

SILVER

**T569, c.1940**
5 tubes, 1 band
brown $75, ivory $135

SILVER

**TL26, c.1940**
4 tubes, 1 band, DC
$60

SILVER

**TS105, c.1940**
5 tubes, 1 band
brown $75, ivory $115

*Silvertone*

**952, c.1931**
9 tubes, 1 band
$75 with speaker

*Silvertone*

**1060/70/80 series, c.1930**
1060/1062 - 6 tubes, 1 band, DC $50
1070/1072/4072/4074 - 7 tubes, 1 band, DC $50
1080/1082/6568/6570 - 8 tubes, 1 band, DC $75
4136/4138 - 9 tubes, 1 band, DC $75

Silvertone
**1140/41, c.1931**
6 tubes, 1 band
$225

Silvertone
**1150/51, c.1931**
6 tubes, 1 band
$300

Silvertone
**1291, c.1932**
6 tubes, 1 band, DC
$250

Silvertone
**1471, c.1932**
6 tubes, 1 band
$225

Silvertone
**1481, c.1932**
7 tubes, 1 band
$225

Silvertone
**1521, c.1941**
6 tubes, 1 band
$40

Silvertone
**1561, c.1941**
6 tubes, 2 bands
$45

Silvertone
**1581, c.1941**
8 tubes, 3 bands
$125

Silvertone
**1591, c.1941**
9 tubes, 4 bands
$225

Silvertone
**1703, c.1934**
4 tubes, 1 band
$75

Silvertone
**1704, c.1934**
5 tubes, 1 band
$125

Silvertone
**1706, c.1934**
5 tubes, 1 band
$225

Silvertone
**1712, c.1934**
8 tubes, 1 band, DC
$150

Silvertone
**1781, c.1941**
8 tubes, 3 bands
$125

Silvertone
**1903, c.1936**
5 tubes, 2 bands
$75

Silvertone
**1761, c.1941**
6 tubes, 2 bands
$100

Silvertone
**1771, c.1941**
7 tubes, 3 bands
$110

Silvertone
**1904, c.1936**
6 tubes, 3 bands
$250

Silvertone
**1905, c.1936**
8 tubes, 3 bands
$225

Silvertone
**1922 'World's Fair', c.1934**
6 tubes, 2 bands, DC
$250

Silvertone
**1923, c.1936**
7 tubes, 3 bands, DC
$175

Silvertone
**1924, c.1936**
6 tubes, 3 bands, DC
$125

Silvertone
**1925, c.1936**
6 tubes, 3 bands, DC
$125

Silvertone
**1928, c.1936**
5 tubes, 1 band, DC
$60

Silvertone
**1938, c.1936**
6 tubes, 3 bands
$110

Silvertone
**1939, c.1936**
5 tubes, 2 bands
$110

Silvertone
**1942, c.1936**
7 tubes, 3 bands
$225

Silvertone
**1947, c.1936**
6 tubes, 3 bands, DC
$125

Silvertone
**2411, c.1941**
5 tubes, 2 bands, DC
$40

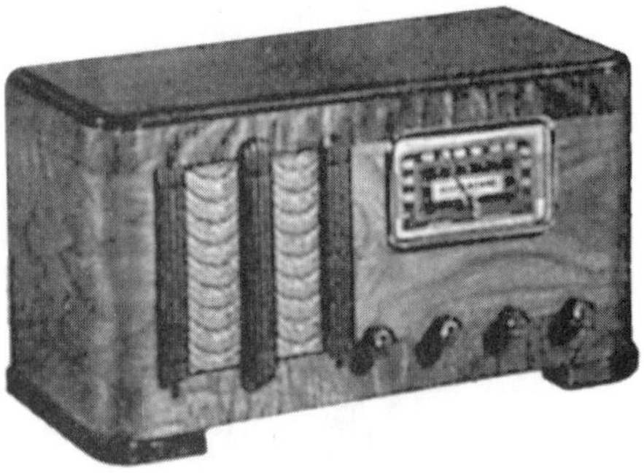

Silvertone
**2421/2721, c.1941**
6 tubes, 1 band, DC
$60

Silvertone
**2441, c.1941**
4 tubes, 1 band, DC
$40

Silvertone
**2451, c.1941**
5 tubes, 1 band, DC
$45

Silvertone
**2461, c.1941**
6 tubes, 3 bands, DC
$45

Silvertone
**2471, c.1941**
7 tubes, 3 bands
$65

Silvertone
**2611, c.1941**
5 tubes, 2 bands, DC
$45

Silvertone
**2641, c.1941**
4 tubes, 1 band, DC
$35

Silvertone
**2651, c.1941**
5 tubes, 1 band, DC
$40

Silvertone
**2861, c.1941**
6 tubes, 3 bands, DC
$60

Silvertone
**3011/3111/3211, c.1941**
5 tubes, 1 band
3011 - brown $100, 3111 - ivory $125
3211 - beetle $225

Silvertone
**3041/3141/3241, c.1941**
4 tubes, 1 band
3041 - black $125, 3141 - ivory $125
3241 - brown $90

Silvertone
**3051/3151/3251, c.1939**
5 tubes, 1 band
3051 - brown $125, 3151 - ivory $150
3251 - beetle $450

Silvertone
**3061/3161/3261, c.1940**
3061/3161/3261 - 6 tubes, 1 band
3361/3461/3561 - 6 tubes, 2 bands
3061/3361 - brown $125
3161/3461 - ivory $150
3261/3561 - beetle $300

Silvertone
**3141/3241/7004,6,8, c.1941**
3141/3241 - 4 tubes, 1 band
7004/6/8 - 5 tubes, 1 band
3141/3004 - brown $125
3142/7006 - ivory $175
7008 - beetle $275

Silvertone
**4614, c.1938**
8 tubes, 3 bands
$150

Silvertone

**4615, c.1938**
6 tubes, 2 bands, DC
$40

Silvertone

**4632/33, c.1938**
6 tubes, 2 bands, DC
$45

Silvertone

**4638/9, c.1939**
8 tubes, 3 bands, DC
$110

Silvertone

**4644/45, c.1938**
6 tubes, 2 bands, DC
$40

Silvertone

**4662, c.1938**
6 tubes, 2 bands
$90

INGRAHAM

Silvertone

**4667, c.1938**
8 tubes, 3 bands
$425

Silvertone

**4675, c.1938**
8 tubes, 3 bands
$300

Silvertone

**4728/9, c.1939**
8 tubes, 3 bands, DC
$75

Silvertone

**4776, c.1939**
6 tubes, 2 bands, radio-phono
$60

Silvertone

**5732/5772, c.1941**
5732 - 8 tubes, 1 band, radio-phono $75
5772 - 8 tubes, 1 band, radio-phono-rec $125

Silvertone

**5741, c.1941**
4 tubes, 1 band, radio, c.19phono
$300

Silvertone

**5751, c.1941**
5 tubes, 1 band, radio, c.19phono
$50

Silvertone

**5771, c.1941**
7 tubes, 1 band, radio-phono
$75

Silvertone

**5811, c.1941**
manual wind phono remote
$75

Silvertone

**5812/5831, c.1941**
5812 - phono remote $75
5831 - wireless phono remote $90

Silvertone

**6228, c.1939**
phono remote
$50

Silvertone

**6028, c.1939**
5 tubes, 1 band, radio-phono
$40

Silvertone

**6050/51, c.1939**
4 tubes, 1 band, DC, radio-phono
$25

**Silvertone**

**6109/10/11 'Rocket', c.1939**
Karstadt Design, 5 tubes, 1 band
6109 brown $1500, 6110 - black $2000
6110 - ivory $2500

**Silvertone**

**6165/67/69, c.1939**
4 tubes, 1 band
6165 - black $150, 6167 - brown $125
6169 - red $600

**Silvertone**

**6188, c.1939**
6 tubes, 2 band, DC
$65

**Silvertone**

**6188, c.1939**
6 tubes, 2 bands, DC
$90

**Silvertone**

**6320, c.1939**
6 tubes, 2 bands
$275

**Silvertone**

**6373, c.1941**
4 tubes, 1 band, DC
$30

**Silvertone**

**6373, c.1941**
4 tubes, 1 band, DC
$60

***Simplex***

**H, c.1931**
5 tubes, 1 band
$325

Simplex

**R, c.1933**
4 tubes, 1 band
$225

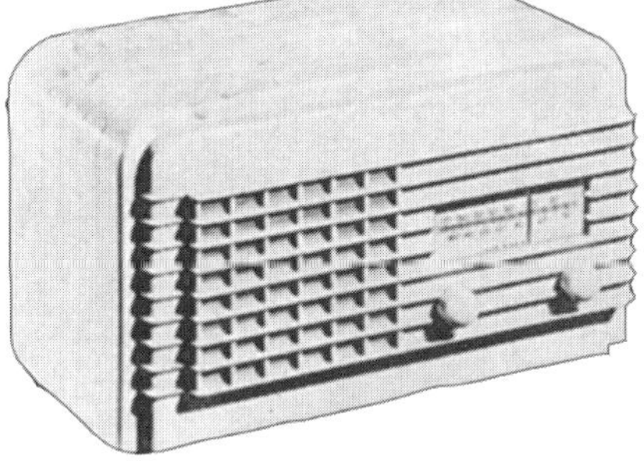

Sonora

**B22, c.1939**
6 tubes, 1 band
$125

Sonora

**DDA14, c.1939**
6 tubes, 2 bands
$60

INGRAHAM

Sonora

**FA55, c.1939**
7 tubes, 3 bands
$110

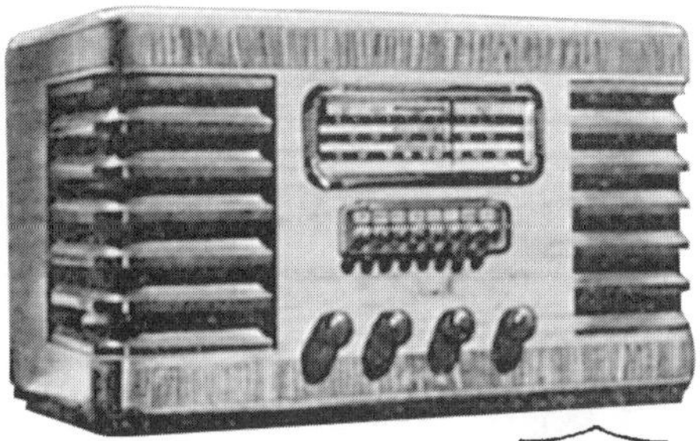

INGRAHAM

Sonora

**GA66, c.1939**
9 tubes, 3 bands
$175

Sonora

**KF, c.1941**
4 tubes, 1 band
combinations of green/black/maroon
$225

Sonora

**KG132, c.1941**
4 tubes, 1 band, DC
$125

Sonora

**KNF99, c.1941**
5 tubes, 1 band, radio-phono
$40

Sonora

**KNF148, c.1941**
5 tubes, 1 band, radio-phono
$55

Sonora

**KT, c.1941**
5 tubes, 1 band
$50

Sonora

**KW78, c.1941**
phono only
$110

Sonora

**KZ111, c.1941**
5 tubes, 1 band, DC
$30

Sonora

**LD93, c.1941**
5 tubes, 2 bands
$80

Sonora

**LLS179, c.1942**
6 tubes, 1 band, AC/DC
$40

Sonora

**LM 'Stratoliner', c.1942**
5 tubes, 1 band, AC/DC
$50

Sonora

**LQ 'Clipper', c.1942**
5 tubes, 1 band, AC/DC
$60

Sonora

**LTF465, c.1942**
5 tubes, 1 band, AC/DC, radio-phono
$125

Sonora

**M22, c.1939**
4 tubes, 1 band, DC
$50

Sonora

**P101 'Jubilee', c.1939**
4 tubes, 1 band
$110

Sonora

**P106 'Metro', c.1939**
4 tubes, 1 band
$90

Sonora

**PLAYETTE JR**
phono remote
$15

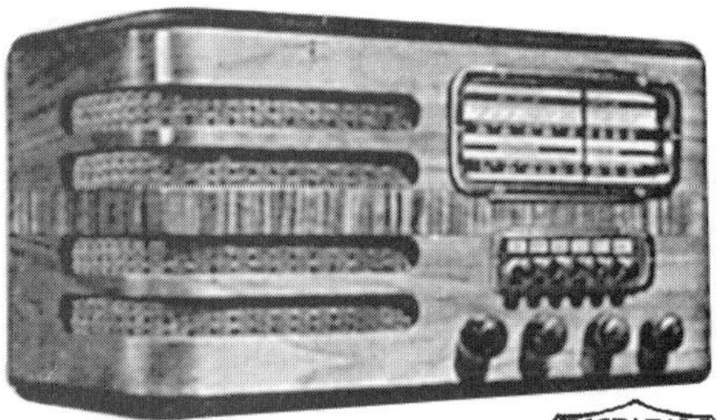

INGRAHAM

Sonora

**QA33, c.1939**
5 tubes, 1 band, DC
$40

INGRAHAM

Sonora

**RA66, c.1939**
8 tubes, 3 bands
$150

INGRAHAM

Sonora

**TE38, c.1939**
phono remote
$150

INGRAHAM

Sonora

**TE41, c.1939**
phono remote
$150

Sonora

**TL26, c.1939**
4 tubes, 1 band, DC
$30

Sonora

**TNE60, c.1940**
6 tubes, 1 band, radio-phono
$75

**Sonora**

**'Troubador', c.1938**
5 tubes, 1 band, radio-phono
$30

**Sonora**

**TT128, c.1940**
5 tubes, 1 band
$150

**SORBER**

**Screen Grid, c.1932**
7 tubes, 1 band, DC
$200

**SPARTON** OF CANADA

**58K, c.1938**
5 tubes, 2 band
$175

**SPARTON** OF CANADA

**58K, c.1938**
Walter Dorwin Teague Attrib.
5 tubes, 2 bands
$275

**SPARTON** OF CANADA

**68K, c.1938**
Walter Dorwin Teague Attrib.
6 tubes, 3 bands
$275

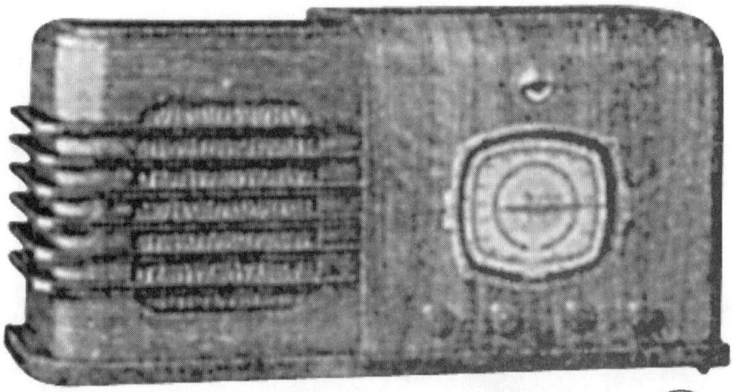

**SPARTON** OF CANADA

**88K, c.1938**
Walter Dorwin Teague Attrib.
8 tubes, 3 bands
$375

Sparton
**62, c.1934**
5 tubes, 1 band
$225

Sparton
**506W, c.1936**
5 tubes, 2 bands
$375

Sparton
**716/716X, c.1936**
7 tubes, 4 bands
$575

Sparton
**Consolette, c.1930**
5 tubes, 1 band
$250

Sparton
**315C 'Football', c.1952**
5 tubes, 1 band, clock-radio
$375

Sparton
**5518A, c.1939**
Walter Dorwin Teague Design
5 tubes, 1 band
$175

## *Steinite*

**Gothic, c.1933**
4/5 tubes, 1 band
$250

STEWART WARNER
CANADA

**Brigadier, c.1938**
4 tubes, 1 band, DC
$50

STEWART WARNER
CANADA

**Grenadier, c.1938**
5 tubes, 2 bands
$125

STEWART WARNER
CANADA

**Little Colonel, c.1938**
5 tubes, 1 band
$90

STEWART WARNER
CANADA

**Mountaineer, c.1938**
6 tubes, 2 bands
$110

INGRAHAM

STEWART WARNER
CANADA

**R467, c.1940**
4 tubes, 1 band, DC
$75

STEWART WARNER

**1-5D9, c.1940**
5 tubes, 1 band, radio-phono
$60

STEWART WARNER

**2A1, c.1941**
phono remote
$50

INGRAHAM

STEWART WARNER

**4B4, c.1940**
5 tubes, 1 band
$225

STEWART WARNER

**6T9, c.1941**
6 tubes, 2 bands
radio-phono-recorder
$100

STEWART WARNER

**7-5R6, c.1940**
5 tubes, 1 band
$150

STEWART WARNER

**7-5R6, c.1940**
7 tubes, 2 bands
$350

STEWART WARNER

**1802, c.1938**
5 tubes, 2 bands
$135

STEWART WARNER

**1881, c.1938**
6 tubes, 2 bands
$60

STEWART WARNER

**1901, c.1938**
5 tubes, 2 bands, DC
$25

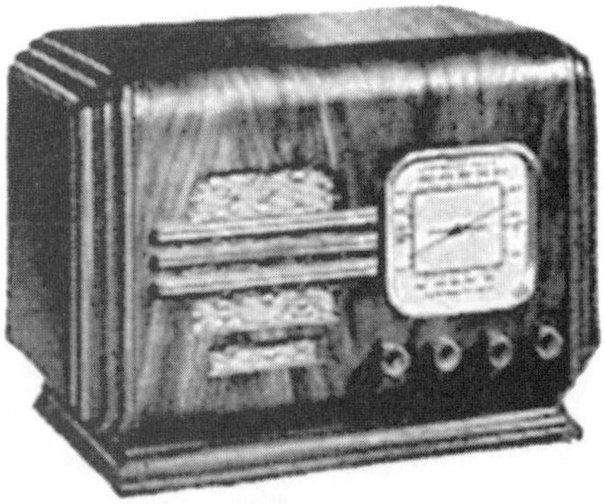

STEWART WARNER

**1911, c.1938**
5 tubes, 1 band, DC
$75

STEWART WARNER

**1921, c.1938**
6 tubes, 2 bands
$200

STEWART WARNER

**Companion Bookset, c.1934**
4 tubes, 1 band
$400

STEWART WARNER

**FM9, c.1941**
9 tubes, FM Converter
$45

STEWART WARNER

**Gulliver, c.1940**
5 tubes, 1 band
$800

STEWART WARNER

**R116, c.1934**
4 tubes, 1 band, 3-position
$600

STEWART WARNER

**R116, c.1934**
4 tubes, 1 band
$1200

STEWART WARNER

**Table, c.1934**
6 tubes, 1 band
$350

Stratford

**Mantel, c.1930**
5 tubes, 1 band
$250

Stromberg-Carlson

**10, c.1933**
4 tubes, 3 bands, SW converter
$150

Stromberg-Carlson

**16, c.1941**
phono remote
$75

Stromberg-Carlson

**411PT, c.1939**
7 tubes, 2 bands, radio/phono
$75

## Stromberg-Carlson

**509PT, c.1941**
5 tubes, 1 band, radio/phono
$75

## Stromberg-Carlson

**520J, c.1941**
7 tubes, 2 bands
$150

## Stromberg-Carlson

**520J, c.1941**
7 tubes, 2 bands, radio/phono
$110

## Stromberg-Carlson

**412H, c.1940**
6 tubes, 3 bands
$125

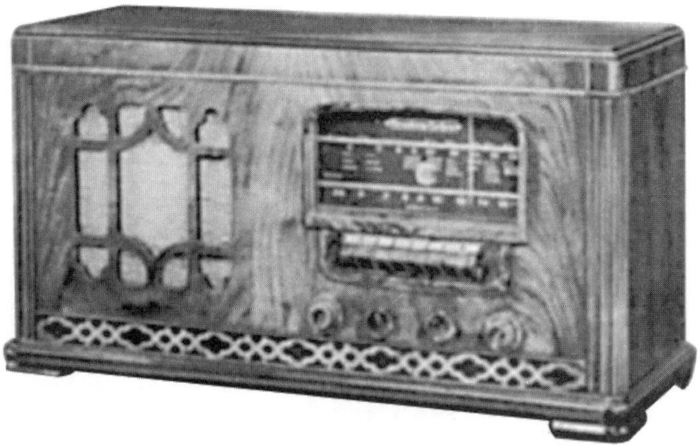

## Stromberg-Carlson

**430H, c.1939**
9 tubes, 2 bands
$200

## Supertone

**26292, c.1935**
5 tubes, 2 bands
$475

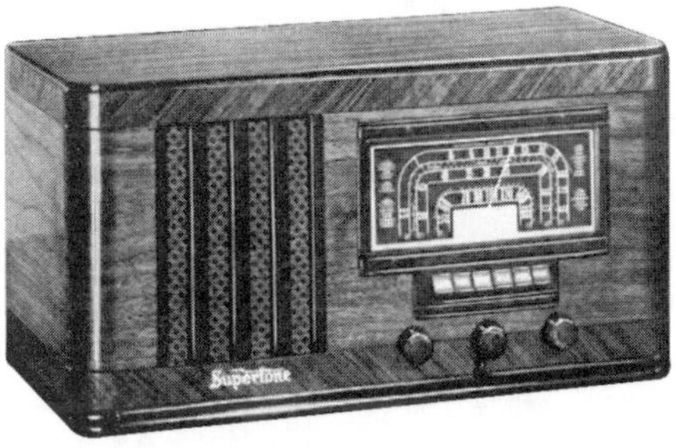

## Supertone

**64182, c.1940**
7 tubes, 2 bands
$75

## Supertone

**64183/4, c.1940**
5 tubes, 1 band, AC/DC
64183 - ivory $175
64184 - brown $175

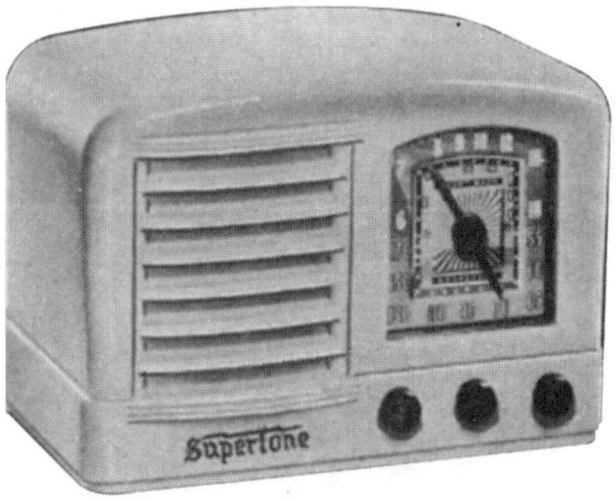


Supertone
**64185/6, c.1940**
6 tubes, 2 bands, AC/DC
64185 - ivory $75
64186 - brown $65

Supertone
**64187/8, c.1940**
5 tubes, 2 bands, AC/DC
64187 - ivory $95
64188 - brown $85

Supertone
**J26294, c.1935**
8 tubes, 3 bands
$275

Supertone
**J26296, c.1935**
5 tubes, 2 bands
$100

Supertone
**J26297, c.1935**
5 tubes, 2 bands
$600

SYLVANIA
**5C10E 'Waltztimer', c.1957**
5 tubes, 1 band, clock-radio
$25

SYLVANIA
**5C11E 'Tempotimer', c.1957**
5 tubes, 1 band, clock-radio
tan $35, blue $50

SYLVANIA
**5C12E 'Niteliter', c.1957**
5 tubes, 1 band, clock-radio
pink $125, aqua $135, tan $75

SYLVANIA
**5C13E 'Skyliter', c.1957**
5 tubes, 1 band, clock-radio
pink $65, blue $65, tan $35

**SYLVANIA**
**5C13E 'Skyliter', c.1959**
5 tubes, 1 band
pink $ 60, tan $35, blue $55

**SYLVANIA**
**5T11E 'Bolero', c.1960**
5 tubes, 1 band
tan $35, blue $50

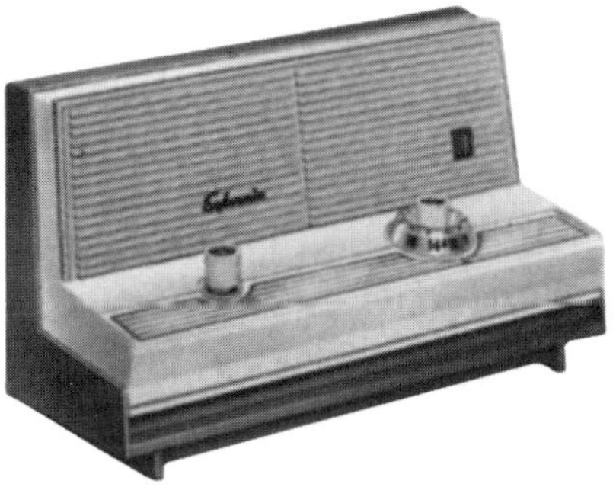

**SYLVANIA**
**5T12E 'Park Avenue', c.1959**
5 tubes, 1 band
tan $75, aqua $125, pink $110

**SYLVANIA**
**5T13E 'Twin Tone', c.1959**
5 tubes, 1 band
pink $75, tan $45, blue $65

**SYLVANIA**
**5T14E 'Music Master', c.1959**
5 tubes, 1 band
pink $90, tan $50, blue $75

**SYLVANIA**
**8F15E 'Recital', c.1960**
transistor, AMFM
$35

**SYLVANIA**
**8F16E 'Interlude', c.1961**
AMFM
$50

**SYLVANIA**
**1107E 'Symphotone', c.1959**
5 tubes, 1 band
black $40

**SYLVANIA**
**2108E 'Swingtimer', c.1959**
5 tubes, 1 band, clock-radio
black $30

**Tele-tone**

**Moderne, c.1946**
5 tubes, 1 band
$75

**Tele-tone**

**Bachelor, c.1946**
5 tubes, 1 band, AC/DC
$225

Tom Thumb

**Junior, c.1934**
5 tubes, 1 band
$1200

**TONE**

**Merry-Go-Sound, c.1948**
phono only
$125

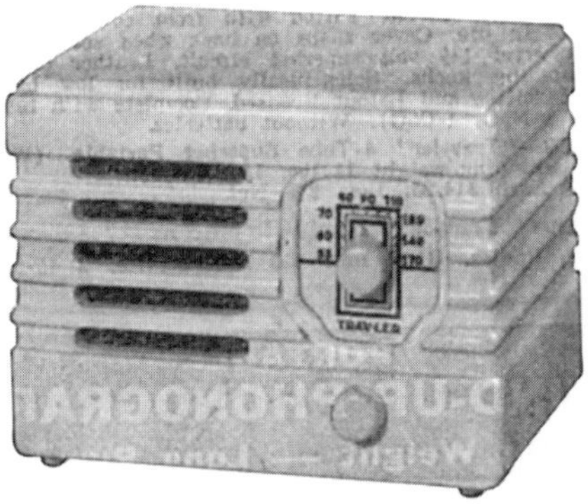

**TRAV-LER**

**400, c.1939**
5 tubes, 1 band
$150

**TRAV-LER**

**41, c.1936**
4 tubes, 1 band
$225

**TRAV-LER**

**440/4/1, c.1938**
4 tubes, 1 band
$55

**TRAV-LER**

**450, c.1938**
5 tubes, 1 band
$125

TRAV-LER

**500, c.1938**
5 tubes, 1 band
$125

TRAV-LER

**539M, c.1940**
8 tubes, 4 bands
$135

TRAV-LER

**626, 'World's Fair', c.1939**
6 tubes, 2 bands
$175

INGRAHAM

TRAV-LER

**c.1936**
5 tubes, 1 band
$500

TRAV-LER

**Midget, c.1931**
5 tubes, 1 band
$250

TROY

**63, c.1935**
5 tubes, 2 bands
$400

TRUETONE

**D2026, c.1952**
8 tubes, AM/FM
$50

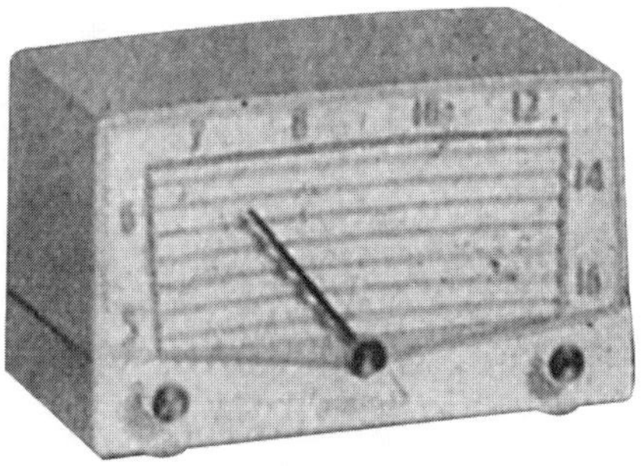

TRUETONE

**D2103, c.1952**
5 tubes, 1 band
$45

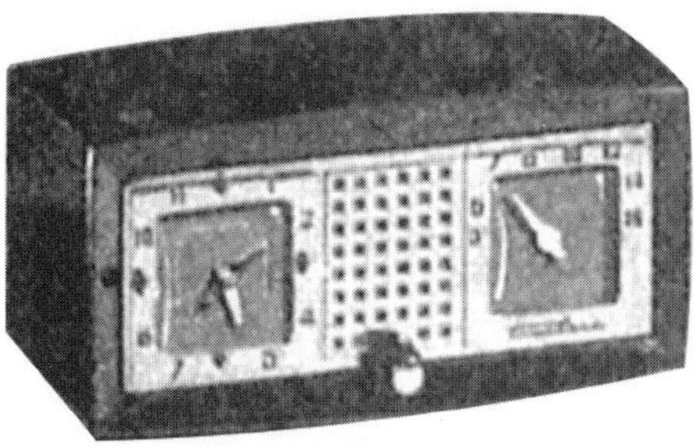

TRUETONE

**D2216, c.1952**
5 tubes, 1 band, clock-radio
$40

**TRUETONE**
**D2237, c.1954**
6 tubes, 1 band
$90

**TRUETONE**
**D2410/11/12/13, c.1954**
5 tubes, 1 band
D2410 - red $110, D2411- green $90
D2412 - ivory $50, D2413 - tan $60

**TRUETONE**
**D2418/19/20, c.1954**
4 tubes, 1 band
D2418 - black $60, D2419 - red $100
D2420 - ivory $45

**TRUETONE**
**D2584/5, c.1954**
5 tubes, 1 band
D2584 - ivory $225
D2585 - brown $175

**TRUETONE**
**D2582, c.1954**
5 tubes, 1 band
$65

**TRUETONE**
**D2637, c.1954**
4 tubes, 1 band
black & pink $125

**TRUETONE**
**D2684/5, c.1954**
5 tubes, 1 band
D2684 - ivory $225
D2685 - brown $175

**TRUETONE**
**D2781/2/3, c.1956**
5 tubes, 1 band
D2781 - red $250, D2782 - $275
D2783 - ivory $125

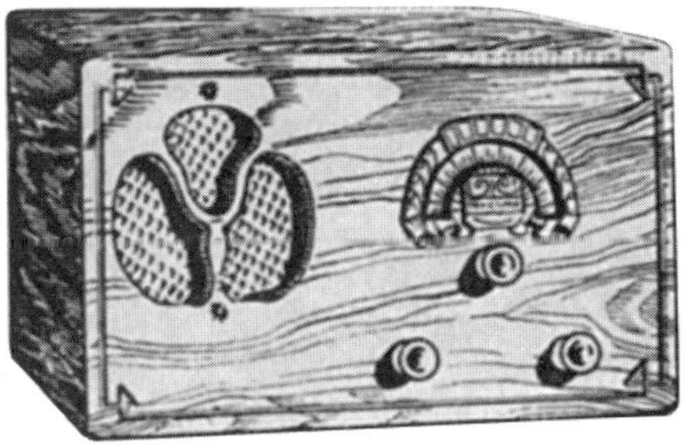

**TRUETONE**
**SLW, c.1933**
5 tubes, 2 bands
$65

**TRUETONE**
**TSS, c.1933**
7 tubes, 2 bands
$175

United Engine
**MIDGET, c.1932**
4 tubes, 1 band
$250

Universal
**72A6/7232, c.1936**
7 tubes, 3 bands, DC
$125

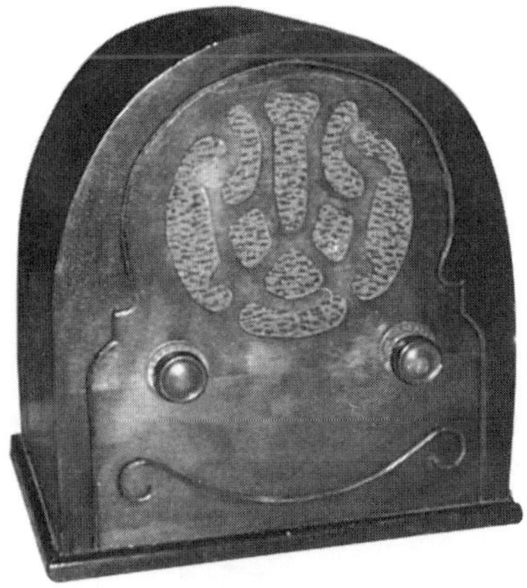

U S RADIO AND REFRIGERATION
**20 'Gloriette', c.1932**
4 tubes, 1 band
$275

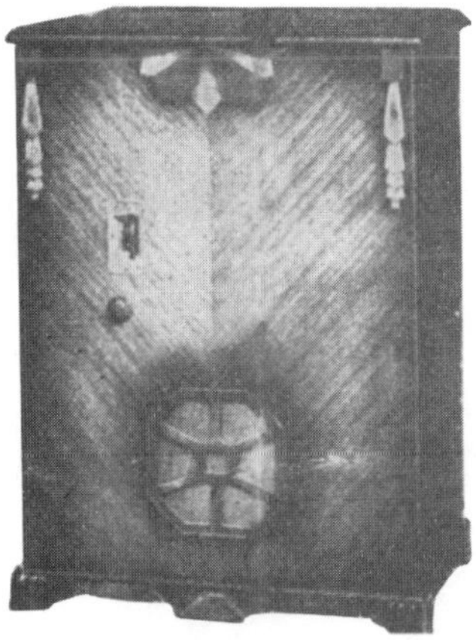

U S RADIO AND REFRIGERATION
**27P, c.1931**
5 tubes, 1 band, radio-phono
$250

WALKER
**Flexi-Unit, c.1932**
1 tube, shortwave adapter
$125

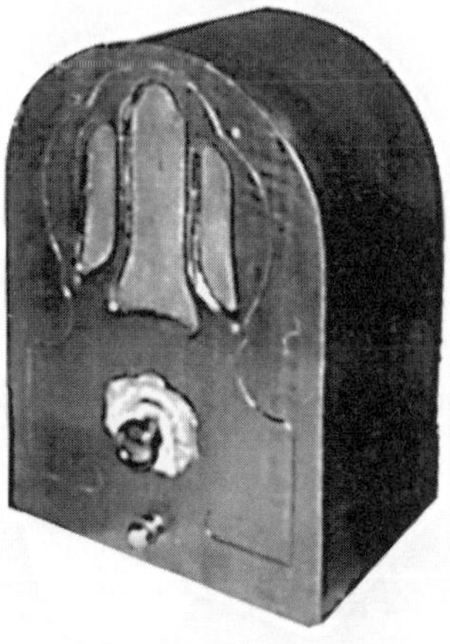

**WALTON**
**Midget, c.1933**
4 tubes, 1 band
$125

WARD
**Police Reporter, c.1934**
1 tube, police band adapter
$50

Westinghouse
CANADA
**420A, c.1938**
4 tubes, 1 band
$225

Westinghouse
CANADA
**527A, c.1938**
5 tubes, 2 bands
$100

Westinghouse
CANADA
**622A, c.1938**
6 tubes, 3 bands
$175

Westinghouse
CANADA
**823A, c.1938**
8 tubes, 3 bands
$200

Westinghouse
CANADA
**555AA, c.1939**
5 tubes, 1 band
$175

WESTRAD
**Junior, c.1930**
6 tubes, 1 band
$200

Westinghouse
**122, c.1947**
5 tubes, 1 band, radio-phono
$125

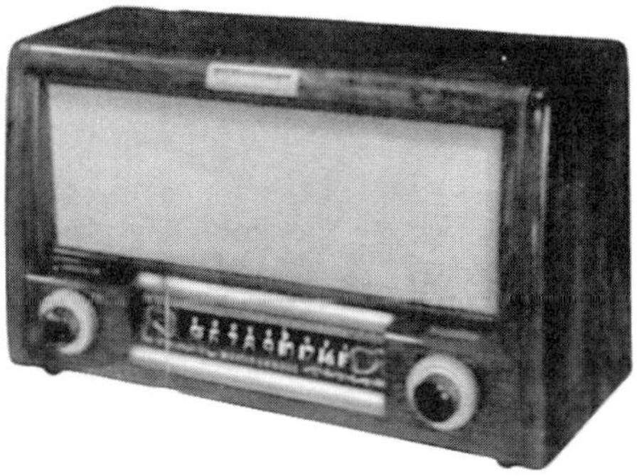

**Westinghouse**

**157, c.1947**
5 tubes, 1 band
$35

**Westinghouse**

**162, c.1947**
8 tubes, AM/FM
$45

**Westinghouse**

**178, c.1947**
4 tubes, 1 band, DC
$35

**Westinghouse**

**H104/5, c.1946**
7 tubes, 2 bands
$50

**Westinghouse**

**H122, c.1946**
6 tubes, 1 band, radio-phono
$125

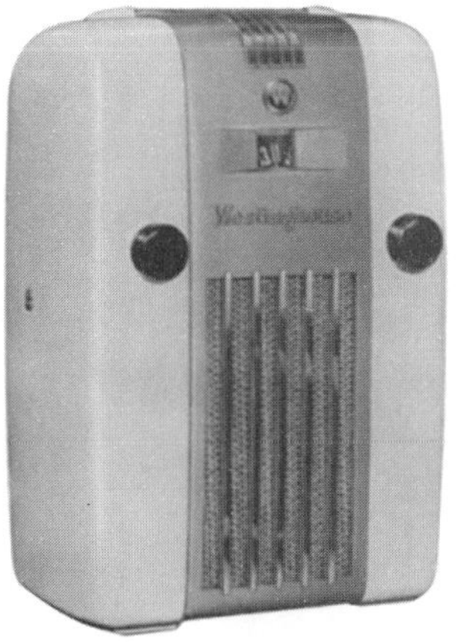

**Westinghouse**

**H125/6/7, c.1946**
6 tubes, 1 bands
H125 - green $175, H126 ivory $150
H127 - maroon $200

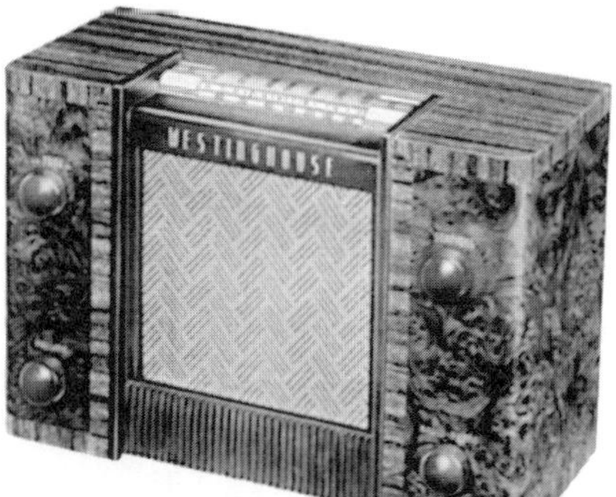

**Westinghouse**

**H130, c.1946**
6 tubes, 1 band
$45

**Westinghouse**

**H133, c.1946**
5 tubes, 1 band
$20

**Westinghouse**

**H157, c.1947**
6 tubes, 1 band
$30

**Westinghouse**

**H161, c.1947**
8 tubes, AM/FM
$45

**Westinghouse**

**H178, c.1948**
5 tubes, 1 band
$30

**Westinghouse**

**H500T5A, c.1948**
5 tubes, 1 band
ivory $75, red $125, brown $50

**Westinghouse**

**WR116, c.1937**
6 tubes, 2 bands
$125

**Westinghouse**

**WR120, c.1938**
6 tubes, 1 band
$85

**Westinghouse**

**WR100, c.1936**
5 tubes, 1 band
$275

**Westinghouse**

**WR10, c.1931**
8 tubes, 1 band
$125

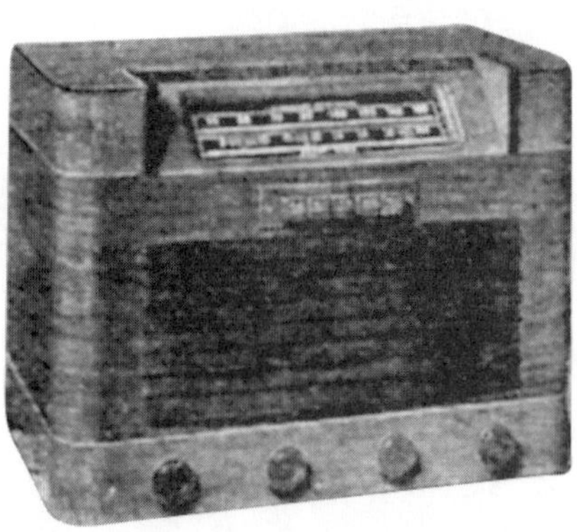

**Westinghouse**

**WR12X16, c.1942**
8 tubes, 2 band, bandspread, AC/DC
$60

**Westinghouse**
**WR12X3, c.1942**
5 tubes, 1 band, AC/DC
$25

**Westinghouse**
**WR12X4, c.1942**
5 tubes, 1 band, AC/DC
$45

**Westinghouse**
**WR14 'Midgette', c.1931**
4 tubes, 1 band
$150

**Westinghouse**
**WR154, c.1939**
6 tubes, 1 band
$45

**Westinghouse**
**WR177, c.1939**
6 tubes, 1 band
$40

**Westinghouse**
**WR182, c.1941**
5 tubes, 2 bands
$35

**Westinghouse**
**WR179, c.1941**
6 tubes, 1 band
$135

**Westinghouse**
**WR184, c.1941**
6 tubes, 1 band
$75

**Westinghouse**
**WR154, c.1939**
6 tubes, 1 band
$60

**Westinghouse**
**WR186, c.1941**
6 tubes, 1 band
$85

**Westinghouse**
**WR203, c.1936**
6 tubes, 3 bands
$235

**Westinghouse**
**WR208, c.1936**
5 tubes, 2 bands
$75

**Westinghouse**
**WR222, c.1936**
5 tubes, 1 band
$75

**Westinghouse**
**WR224, c.1938**
5 tubes, 2 bands
$110

**Westinghouse**
**WR228, c.1938**
8 tubes, 3 bands
$150

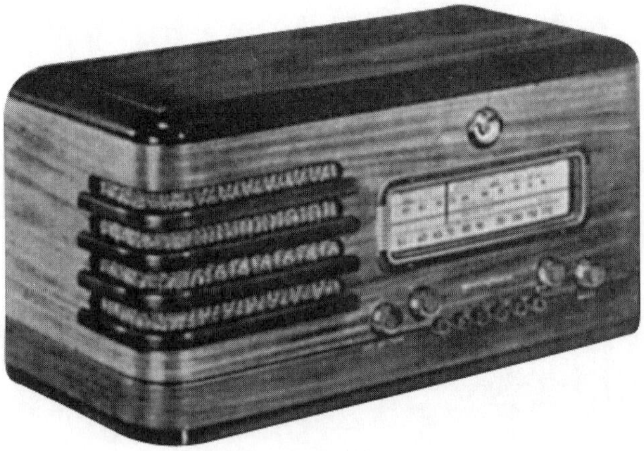

**Westinghouse**
**WR262, c.1939**
6 tubes, 2 bands
$125

**Westinghouse**

**WR264, c.1939**
7 tubes, 2 bands
$70

**Westinghouse**

**WR288, c.1941**
8 tubes, 3 bands
$125

**Westinghouse**

**WR472, c.1939**
6 tubes, 1 band, radio-phono
$40

**Westinghouse**

**WR478, c.1941**
5 tubes, 1 band, radio-phono
$30

**Westinghouse**

**WR602, c.1936**
6 tubes, 2 bands, DC
$65

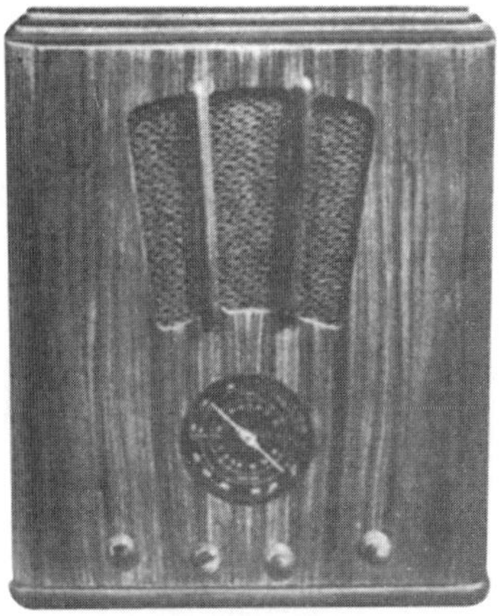

**Westinghouse**

**WR604, c.1936**
6 tubes, 2 bands, DC
$45

**Wilcox-Gay**

**A56, c.1939**
phono remote
$25

**Wilcox-Gay**

**A57, c.1939**
phono remote
$35

**Wilcox-Gay**

**A63, c.1939**
phono remote
$75

Wilcox-Gay

**Cardinal, c.1934**
7 tubes, 2 bands
$1500

**WURLITZER**

**Eagle, c.1934**
7 tubes, 2 bands
$750

WORLD

**SW, c.1934**
1 tube, shortwave
$75

WORLD

**Troubador, c.1936**
5 tubes, 2 bands
$425

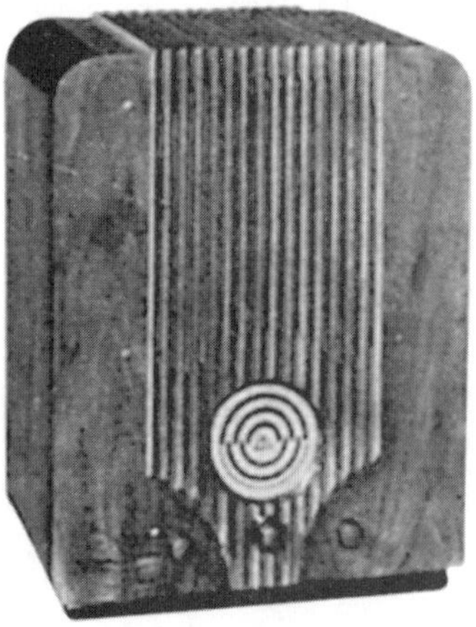

WORLD

**583/4, c.1934**
5 tubes, 2 bands
$600

ZENITH

**6D-302, c.1939**
6 tubes, 1 band, radio-phono
$350

ZENITH

**5R-303, c.1939**
6 tubes, 1 band, radio-phono
$225

ZENITH

**6S-306, c.1939**
6 tubes, 2 bands, radio-phono
$350

ZENITH

**4B-313, c.1939**
4 tubes, 1 band, DC
$150

ZENITH

**7J-328, c.1939**
7 tubes, 3 bands, AC/DC
$375

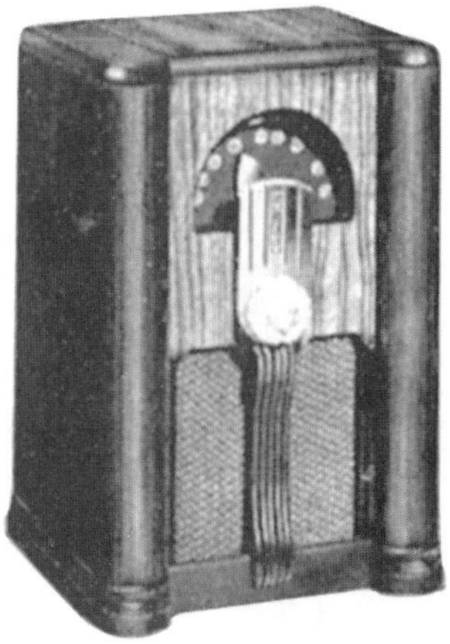

ZENITH

**6D-326 'Child's Radio', c.1939**
6 tubes, 1 band
$900

ZENITH

**5R-327, c.1939**
5 tubes, 2 bands
$500

ZENITH

**6S-330, c.1939**
6 tubes, 2 bands
$250

ZENITH

**403, c.1940**
5 tubes, 1 band, AC/DC
$125

ZENITH

**437, c.1940**
4 tubes, 1 band, DC
$75

ZENITH

**480, c.1940**
6 tubes, 1 band, radio-phono
$300

ZENITH

**481, c.1940**
6 tubes, 1 band, radio-phono
$425

ZENITH

**580, c.1941**
6 tubes, 2 bands, radio-phono
$150

ZENITH

**583, c.1941**
6 tubes, 2 bands, radio-phono
$125

ZENITH

**616, c.1942**
5 tubes, 1 band
$90

ZENITH

**617, c.1942**
4 tubes, 1 band, DC
$50

ZENITH

**635, c.1942**
4 tubes, 1 band, DC
$75

ZENITH

**636, c.1942**
5 tubes, 1 band
$110

Zenith

**636, c.1942**
5 tubes, 1 band
$110

Zenith

**638, c.1942**
6 tubes, 1 band
$90

Zenith

**639, c.1942**
4 tubes, 1 band, DC
$50

Zenith

**640, c.1942**
4 tubes, 1 band
$60

Zenith

**680, c.1942**
5 tubes, 1 band, radio-phono
$75

Zenith

**683, c.1942**
6 tubes, 1 band, radio-phono
$90

Zenith

**684, c.1942**
6 tubes, 1 band, radio-phono
$75

Zenith

**686, c.1942**
5 tubes, 1 band, radio-phono
$75

Zenith

**7000, c.1940**
phono only
$40

ZENITH

**7002, c.1940**
phono remote
$110

ZENITH

**8500, c.1940**
wireless phono remote
$40

ZENITH

**8501, c.1941**
wireless phono remote
$60

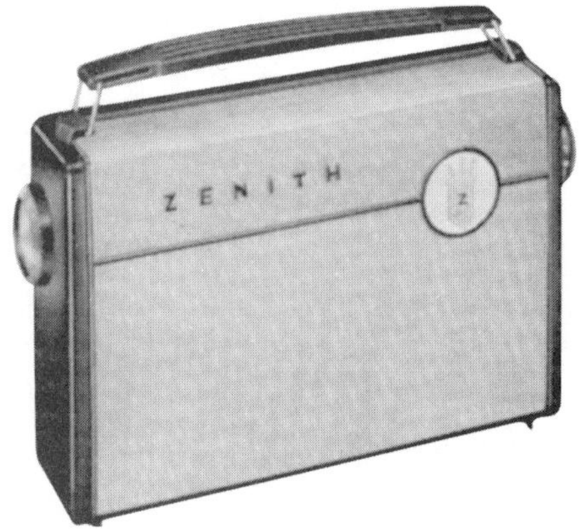


ZENITH

**B400, 'Holiday', c.1958**
5 tubes, 1 band
black $30, aqua $50, pink $50

ZENITH

**A402 'Seashore', c.1958**
5 tubes, 1 band
black $40, tan $45, pink $50

ZENITH

**A501 'Laurel', c.1959**
5 tubes, 1 band
black $40, white $40, green $75

ZENITH

**A515L, c.1959**
5 tubes, 1 band, clock-radio
tan $50, black $50, green $75, white $60, black & red $90

ZENITH

**A516W 'Deluxe', c.1959**
5 tubes, 1 band, clock-radio
white $40, tan $50, grey $60

ZENITH

**A 'Zenette', c.1931**
6 tubes, 1 band
$350

ZENITH

**AH 'Zenette', c.1931**
6 tubes, 1 band
$350

ZENITH

**B 'Zenette', c.1931**
6 tubes, 1 band
$600

ZENITH

**L 'Zenette', c.1931**
5 tubes, 1 band
$400

ZENITH

**B508 'Majorette', c.1959**
5 tubes, 1 band
white $35, maroon $45

ZENITH

**B510 'Lancer', c.1959**
5 tubes, 1 band
black $45

ZENITH

**B511 'Trumpeteer', c.1959**
5 tubes, 1 band
tan $40, aqua $65, yellow $75, orange $75

ZENITH

**B513 'Toreodor', c.1959**
5 tubes, 1 band
green $65, black $50, orange $75

ZENITH

**B514, c.1959**
5 tubes, 1 band, clock-radio
pink $75, white $45, green $90, grey $60

**B515, c.1959**
5 tubes, 1 band, clock-radio
black $50, blue $90, yellow $90, grey $60

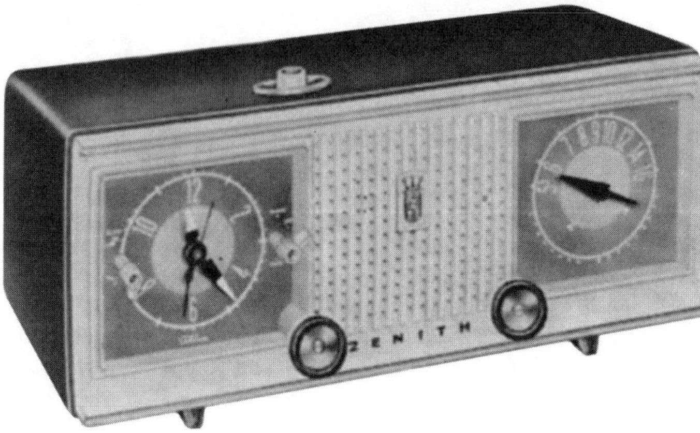


**C519 'Nocturne', c.1959**
5 tubes, 1 band, clock-radio
black $50, tan $50, yellow $90, white $50

**Golden Triangle, c.1960**
transistor, clock-radio
$200

**K666 'Malibar', c.1953**
6 tubes, 1 band, radio-phono
$90

**B509 Ascot, c.1959**
5 tubes, 1 band
green $50, yellow $60, red $70, grey $50

**K515, c.1953**
5 tubes, 1 band, clock-radio
brown $75, black $100

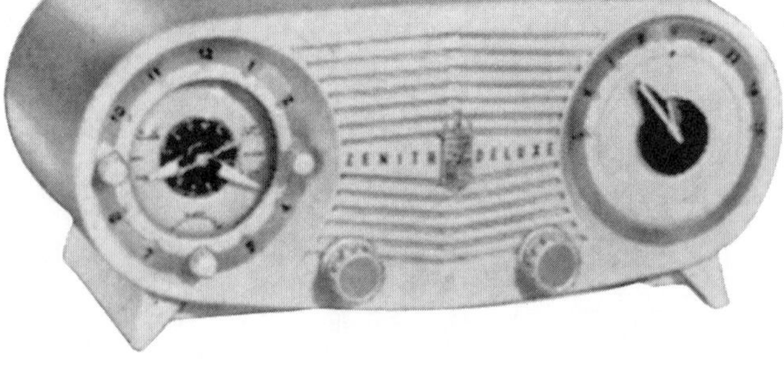


**K518, c.1953**
5 tubes, 1 band, clock-radio
$100

ZENITH

**K622, c.1953**
6 tubes, 1 band, clock-radio
$75

ZENITH

**Y514, c.1959**
6 tubes, 1 band, clock-radio
brown $75, ivory $100, blue $115

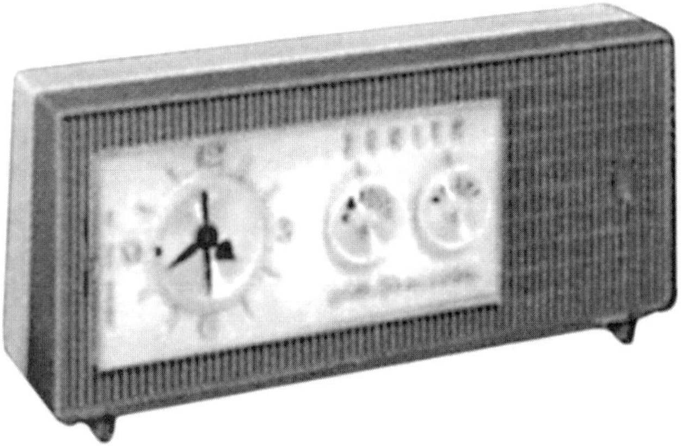

ZENITH

**ROYAL850, c.1957**
transistor, 1 band, DC, clock-radio
black $40, ivory $35, tan $45, pink $60

ZENITH

**Z615F 'Zephyr', c.1958**
5 tubes, 1 band
white $45, grey $50, green $65

ZENITH

**B615 'Cotillion', c.1957**
5 tubes, 1 band
yellow $75, green $60, grey $50

ZENITH

**S9001, c.1941**
phono remote
$30

ZENITH

**S9002, c.1941**
phono remote
$30